KEY NOTES ON AGRICULTURE CHEMISTRY AND SOIL SCIENCE

For Ready Reference to the

STUDENTS, TEACHERS, RESEARCHERS & ASPIRANTS OF COMPETITIVE EXAMINATIONS

THE EDITORS

Dr. U.D. Chavan obtained his M.Sc. (Agri. in Biochemistry) degree from Mahatma Phule Krishi Vidyapeeth, Rahuri. He received his Ph.D. degree in Food Science from Memorial University of Newfoundland St. John's Canada in 1999. He has done International Training on "Global Nutrition 2002" at Uppsala University Uppasala, Sweden in 2002. Dr. Chavan worked as Senior Research Assistant in the Department of Biochemistry & Food Science and Technology at MPKV Rahuri from 1988 to 2000. During his Ph.D., he worked as Technician/Research Associate at Atlantic Cool Climate Crop Research Center and Agriculture and Agri-Food Canada. He received D.Sc. degree in 2006 from USA.

Dr. Chavan is presently working as a Senior Cereal Food Technologist in the Department of Food Science & Technology at Mahatma Phule Krishi Vidyapeeth, Rahuri.

Dr. J.V. Patil obtained his M.Sc. (Agri.) from, MPKV, Rahuri. He completed his course work for Ph.D. at CCSHAU, Hisar and research at MPKV, Rahuri in 1992. He rendered his research and teaching services at MPKV Rahuri as Geneticist, Associate Professor, Plant Breeder and Professor of Genetics & Plant Breeding and Head, Genetics and Plant Breeding Department, MPKV, Rahuri. He also delivered many administrative responsibilities in the University. Dr. Patil joined as the Director, Directorate of Sorghum Research, Hyderabad in August 2010.

THE CONTRIBUTORS

Dr. R.U. Nimbalkar is Junior Research Assistant in the Department of Soil Science & Agril. Chemistry at Mahatma Phule Krishi Vidyapeeth, Rahuri.

Dr. A.D. Kadlag is an Associate Professor in the Department. of Soil Science & Agril. Chemistry at Mahatma Phule Krishi Vidyapeeth, Rahuri.

KEY NOTES ON AGRICULTURE CHEMISTRY AND SOIL SCIENCE

For Ready Reference to the

STUDENTS, TEACHERS, RESEARCHERS & ASPIRANTS OF COMPETITIVE EXAMINATIONS

Editors

U.D. CHAVAN

&

J.V. PATIL

Contributors

R.U. NIMBALKAR

A.D. KADLAG

2015

Daya Publishing House®

A Division of

Astral International (P) Ltd

New Delhi 110 002

© 2015 PUBLISHER
ISBN: 9789351306993 (International Edition)

Published by	:	**Daya Publishing House®**
		A Division of
		Astral International Pvt. Ltd.
		– ISO 9001:2008 Certified Company –
		4760-61/23, Ansari Road, Darya Ganj
		New Delhi-110 002
		Ph. 011-43549197, 23278134
		E-mail: info@astralint.com
		Website: www.astralint.com
Laser Typesetting	:	**Twinkle Graphics, Delhi**
Printed at	:	**Thomson Press India Limited**

PRINTED IN INDIA

PREFACE

India is an agricultural country. The Indian economy is basically agarian. Inspite of economic and industrialization, agriculture is the backbone of the Indian economy. As Mahatma Gandhi said "India's lives in villages and agriculture is the soul of Indian economy". Agriculture is a vast subject and encompasses at least 20 major and minor subjects in it. New developments have lead to entirely a new face of agriculture. Study of agriculture has always been intrigued with a mosaic of interwove concepts, subjects, facts and figures. There are number of books and large literature on Agriculture Chemistry and Soil Science but the Key Notes type of book have not been compiled in a readable manner.

The present book *"Key Notes on Agriculture Chemistry and Soil Science"* has been designed to fulfill this long felt need of students, teachers, researchers and aspirants of competitive examinations. It is designed in such a way that give rapid, easy access to the core materials in a short format which facilitates easily learning and rapid revision. The book carries fundamentals of Agriculture Chemistry and Soil Science. There are 7 chapters elaborating Discoveries, Abbreviations, Terminology, Reasoning: Principles and Concepts, Explanations, Diagrammatic Representation and Short Notes, Formulae and Equations as well as references also included. The most recent information is provided along with a detailed list of references for further reading.

Hope this book would be highly useful for graduate and post-graduate students of agriculture, teachers and researchers. This book will also useful for the aspirants of various competitive examinations such as Agricultural Research Service (ARS), ICAR- National Eligibility Test (NET), State Eligibility Test (SET), Junior Research Fellowship (JRF), Senior Research Fellowship (SRF), Civil Services, Allied Agricultural Examinations and Extension Workers for reference and easy answers of many complicated questions. Thus it is expected that this book will adequately meet the need of wider circle of students and readers for preparing their professional career.

We acknowledge the references that are used in this manuscript. Authors are also thankful to all scientists and friends who have helped directly or indirectly while preparing this manuscript. The editors of grateful to all the contributors for their cooperation, support and timely submission of their manuscripts for bringing out this publication. We would have like to acknowledge the patience and support of our families whilst we have spent many hours with drafts of manuscripts rather than with them. Lastly, our sincere thanks to publisher Astral International Pvt. Ltd., New Delhi who provides an opportunity to publish this book.

To all readers we extend an invitation to report that no doubts have escaped our attention and to offer suggestion for improvements that can be incorporated in future editions.

U.D.Chavan and J.V. Patil

Editors

CONTENTS

1

DISCOVERIES

DISCOVERIES

Sr.No.	Discoveries	Scientist	Year
1.	Law of Limiting Factor	Blakman	
2.	(a) Electrochemical Theory of Nutrient Uptake	Lundergardh	
	(b) Theory of Relativity	Lundergardh	
3.	Electroultrafiltration Theory	Nyrnith (Q/T)	
4.	Target Yield Concept	Ramamorthy et al.	
5.	(a) Equilibrium Phosphate Potential	Ramamoorthy –	
	(b) Phosphate Equilibrium Potential	Subramanian	
6.	'E' value	Russell	
7.	Soil is an inexhaustible source of nutrients	Cameroon	
8.	Per cent Yield Concept	Baule	
9.	(a) Nutrient Mobility Concept		
	(b) Per cent Sufficiency Concept	Bray Bray	
10.	Law of Minimum/Law of Limiting Nutrient or Law of Restitution	Liebig	1862
11.	Critical Soil Test Level Concept	Cate and Nelson	
12.	Benchmark Soils of India	Murthy et al.	
13.	Ultramicronutrient Concept (e.g., Co)	Nicholas	
14.	(a) CEC if next to Photosynthesis	Marshall	
	(b) pH is physical phenomenon	Marshall	
15.	Carrier Theory	Epstein	
16.	Biogas (cattle dumg prepn)	S.V. Desai and S.C. Biswas	
17.	Autoradiography	La cassageue	
18.	Nutrient Sufficiency concept	Baule	
19.	Lignin Theory	Waksman	
20.	Water – essential for plant	Van Hement	
21.	Water – transpiration	Woodward	
22.	Ionic strength	Lewis and Randol	
23.	Isotopic Dilution	Hevesy and Hoffer	
24.	Auto Radioactivities	Reinganum	
25.	C.E.C.	Thompson and Way	1977
26.	Root exchange (concept)	Jenny and Overstreet	

Contd...

Sr.No.	Discoveries	Scientist	Year
27.	Root C.E.C.	Devaux	
28.	pH concept	Sorenson	1909
29.	pF concept	Silon	
30.	Prescription Approach	Troug	
31.	Father of Soil Testing	Troug	1954
32.	Integrated Soil Test Approach	Colwell	
33.	DRIS concept	Beaufils	
34.	Radio Carbon Dating	Libby et al.	
35.	Lime potential	Aslyng and latter Schofield	
36.	Q/I	Beckett	
37.	Salt Index (SI)	Puri	
38.	Soil Catena	Milne	
39.	Polyphenol Theory of Organic Matter	Stevenson	
40.	Fertility/Nutrient Index	Parker	
41.	DTPA – extractable Micronutrients	Lindsay and Norvell	1978
42.	'A' value	Fried and Dean	
43.	'Y' or 'L' value	Larsen	
44.	Mathematical equation indicating relationship between plant growth response and growth factor	Mitscherlich	1909
45.	Phenomenon of radioactivity	Becquerel	1896
46.	Liquid Scintillation counting	Kallman and Reynold	
47.	Azomethine – H method for determination of available boron (B) in soils and irrigation water	Berger and Truog	1939
48.	Carmine method for determination of boron (B) in soil and irrigation water	Hatcher and Wilcox	1950
49.	Bicarbonate extraction method for determination of available sulphur in soils	Tabatabai	1982
50.	Monocalcium Phosphate Monohydrate Extraction method for determination of available sulphate – 5 of soil	Schulte and Eik	1988
51.	Determination of exchangeable cations of soil (Ca, Mg, Na)	Tucker	1960
52.	Cation Exchange Capacity (CEC) of soil by $Mg(NO_3)_2.6H_2O$ method	Polemio and Rhoades	1977
53.	Adama and Evans and methods for lime requirement of acid soil	Mclean	1982
54.	Wet oxidation method for determination of organic carbon from soil	Nelson and Sommers	1982
55.	Soil testing for available phosphorus	Olsen and Sommers	1982
56.	Determination of available 'K' from soil	Knudsen and Peterson	1982
57.	Determination of alkaline earth carbonates of soil	Nelson	1982
58.	Determination of calcium and magnesium carbonates in soils	Mahi et al.	1987
59.	Measurement of soil salinity	Dahnke & Whitney	1983
60.	Sulphur Cycle	Trundinger	1979

SCIENTISTS INVOLVED IN VARIOUS DISCOVERIES

Year/Period	Name of scientist	Discovery/Contribution
1872-1919	Mikhail Tsvet	Invented Chromatography, research on plant pigments
1871-1937	Ernest Rutherford	The concept of radioactive half life and discovery of proton
1867-1934	Marie Curie	Pioneer of Radioactivity and the first person honored with two Nobel Prizes
1875-1946	Gilbert Newton Lewis	Discovery of the covalent bond, Lewis dot structures
1834-1907	Dmitri Mendeleev	Creator of the first version of the periodic table of elements
1745-1827	Alessandro Giuseppe Antonio Anastasio Volta	Discovered Methane by collecting the gas from marshes
1852-1911	Jacobus Henricus van't Hoff	Discoveries in chemical kinetics, chemical equilibrium, osmotic pressure and stereochemistry
1733-1804	Joseph Priestley	Discovery of oxygen, having isolated it in its gaseous state
1779-1848	Jons Jacob Berzelius	Modern technique of chemical formula notation (law of constant proportions) and known as father of modern chemistry
1829-1896	Friedrich August Kekule von Stradonitz	Founder of the theory of chemical structure
1901-1994	Linus Carl Pauling	The nature of chemical bonds and the structure of molecules
1853-1932	Wilhelm Ostwald	Electrochemistry and electrolytic dissociation of organic acids, invented viscometer

2

ABBREVIATIONS

Abbreviation	Full forms
AEC	Anion Exchange Capacity
AFS	Apparent Free Space
AI	Activity Index
Alf	Alfisol
And	Andisol
API	Aerial–photo–interpretation
APS	Adenosine phospho sulphate
$AR^k e$	Activity ratio of k at equilibrium
ATP	Adenosine triphosphate
b	biosphere
B.D.	Bulk Density
BGA	Blue Green Algae
BHC	Benzene hexachloride
BOD	Biological Oxygen Demand
BS	Base Saturation
C:N Ratio	Carbon: Nitrogen Ratio
CCC	Chlorocholine Chloride
CCE	Calcium Carbonate Equivalent
CDU	Crotonylidene diurea
CEC	Cation Exchange Capacity
Cl	Climate
CNC	Critical Nutrient Concentration
COD	Chemical Oxygen Demand
d.p.s.	disintegrations per second

Abbreviation	Full forms
DDL	Diffuse Double Layer
DFS	Donnan Free Space
DTPA	Diethylene Triamine Pent Acetic acid
EC	Electrical Conductivity
ECC	Effective Calcium Carbonate
EDTA	Ethylene Diamine Tetra Acetic acid
Eh	Redox potential
E_{max}	Maximum energy
EMR	Electromagnetic Radiation
Ent	Entisol
EPP	Equilibrium Phosphate Potential
ESP	Exchangeable Sodium Percentage
ET	Evapotranspiration
EUF	Electro-ultra filtration
Ew	Erosion due to action of water
f	Function
FYM	Farm Yard Manure
GIS	Geographic Information System
GMD	Geometric Mean Diameter
GR	Gypsum Requirement
H.C.	Hydraulic Conductivity
H.I.	Harvest Index
IBDU	Isobutylidene diurea
IUSS	International Union of Soil Science
Ksp	Specific sites for potassium
LAI	Leaf Area Index
LCU	Lac Coated Urea
LR	Leaching Requirement
MeV	million electron volt
MWD	Mean Weight Diameter

Abbreviation	Full forms
N.I.	Neutralizing Index
N.V.	Neutralizing Value
NADP	Nicotinamide Diphosphate
NAR	Net Assimilation Rate
NCSS	National Co-operative Soil Survey
NCU	Neem Coated Urea
NiR	Nitrite Reductase (enzyme)
NR	Nitrate Reductase (enzymes)
O	Organisms
O.C.	Organic Carbon
O.M.	Organic Matter
O.P.	Osmotic Pressure
ODR	Oxygen Diffusion Rate
Ox	Oxisol
p	Parent materials
P.D.	Particle Density
PAPS	Phosphoadenine Phospho Sulphate
PBC^k	Potential buffering capacity of potassium
pdff	phosphorus in the plant derived from fertilizer
PEP	Phosphoenol pyruvate
Pi	Inorganic phosphate
PTE_S	Potentially toxic elements
pzc	point of zero charge
r	relief
rd	Rutherford
R_{max}	Maximum range
RSC	Residual Sodium Carbonate
S	Soil formation
SAR	Sodium Adsorption Ratio
SCU	Sulphur Coated urea

Abbreviation	Full forms
SI	Stability Index
Sp. Heat	Specific Heat
SPAC	Soil-Plant-Atmosphere-Continuum
SS	Structural Co-efficient of Soils
SSP	Soluble Sodium Percentage
t	time
TPP	Thiamine pyrophosphate
UF	Urea Formaldehyde
Ult	Ultisol
USDA	United States Department of Agriculture
USLE	Universal Soil Loss Equation
VAM	Vesicular Arbuscular Michorhizae
WFS	Water Free Space
WUE	Water Use Efficiency
α	Alpha
β	Beta
γ	Gamma

3

TERMINOLOGY

Term	Definition
'A' Value	The amount of available nutrient in soil to be determined in terms of fertilizer standard is known as 'A' value.
'L' value	Is the amount of phosphorus in the soil and in the soil solution which is exchangeable with added orthophosphate to the soil as measured by a plant growing in the system?
Acid rain	Is in fact a natural phenomenon occurring during any thunderstorm accompanied by heavy lightening or by volcanic eruptions.
Active acidity	May be defined as the acidity develops due to hydrogen (H^+) and aluminium (Al^{3+}) ions concentration of the soil colloids. The magnitude of this exchange acidity is very high.
Active transport	Is the process by which ion cross the root from the epidermis to the xylem with the expenditure of metabolic energy against the gradient of their concentration?
Adhesive bonding	The attraction of unlike molecules for each other is adhesive bonding.
Adsorption	Is defined as a phenomenon by which an increase in concentration or an accumulation of an ion species on a solid occurs due to ion exchange or other reactions.
Alpha (α) particles	Are doubly positively charged helium atoms.
Aminisation	The process by which the hydrolytic decomposition of proteins from combined nitrogenous compounds as well as release of amines and amino acids takes place by heterotrophic soil micro-organisms. Heterotrophic micro-organisms Proteins R-NH + CO_2 + energy + other additional products (combined with organic materials)

Term	Definition
Ammonification	In this process organic nitrogenous compounds transform to NH_4^+ FORM (inorganic) by enzymic hydrolysis through some intermediate steps as follows: Organic nitrogenous compounds polypeptides amino acids NH_3 or salts of NH_4^+ e.g., proteins.
Argillation	Is a process by which the dispersal clay particles are migrated from upper to the lower soil horizon resulting a textural horizon?
Atomic or nuclear energy	In nuclear fission and fusing reactions the average binding energy per nucleon is greater in the product nuclei than in the reacting nuclei and hence, energy is released in such nuclear reactions and is known as atomic or nuclear energy.
Available water	That portion of water which is retained in the soul between field capacity ($-1/3$ bar) and the permanent wilting coefficients (-15 bars).
Azotobacterium	The bacterial fertilizer of *Azotobacter chrococcum* is known as "Azotobacterin".
Bar or atmosphere	Atmosphere or bar is the average air pressure at sea level.
Beta (β) particles	Are electrons emitted by the unstable nuclei of the radioactive atoms at the time of its disintegration or decay?
Bio-fertilizers or Microbial inoculants	Are defined as preparations containing live or latent cells of efficient strains of nitrogen fixing, phosphate solubalizing or cellulolytic micro-organisms used for application to seed, soil or composting areas with the objectives of increasing the population of such beneficial micro-organisms and accelerate certain microbial processes to augment the extent of the availability of nutrients in a firm which can be easily assimilated by plants.
Biological nitrogen fixation	Nitrogen from inorganic molecular form in the atmosphere is fixed and converted to an organic form by a number of various kind of soil micro-organisms is known as biological nitrogen fixation.

Term	Definition
Biological weathering	Disintegration and decomposition of rock is and minerals by man, and animal, root of higher plants, different micro organisms etc.
Bremsstrahlung	When beta particles approaches towards the nucleus of an atom, the speed of the particles becomes slow resulting from the loss of energy and as a result continuous emission of X-rays occurs and of K, L or M orbits of any atom, this is known as "Bremsstrahlung".
Buffer Action	The power to resist a change in p^H is called buffer action.
Bulk density C:N Ratio	Is defined as mass (weight) per unit volume of a dry soil (volume of solid and pore spaces). The nitrogen content in the microbes and in the organic residues is given in proportion to the carbon content and that is called Carbon:Nitrogen.
Calcification and Gypsification	Are the forming processes of arid and semi-arid regions and refer to the formation and accumulation of calcium refers to the formation and accumulation of calcium carbonate and gypsum respectively.
Capillary water	May be defined as the water that is retained in the soil between the water potential of $-1/3$ bar to -31 bars.
Carbonation	Carbonation is the combination of carbon dioxide with any base.
Cation exchange	The process of interchange of cations in soil solution with those in exchangeable form is known as cation exchange.
Cation Exchange Capacity (CEC)	Is the amount of exchangeable cations per unit weight of dry soil?
Cellulose	Is a carbohydrate composed of glucose units bound together in a long, linear chain by linkages at carbon atoms 1 and 4 of the sugar molecule?
Chelate	Is defined as type of chemical compound in which a metallic ion is firmly combined with a molecule by means of multiple chemical bonds.
Chemical weathering	May be defined as the transformation of original rocks and minerals into new compounds having different chemical composition and physical properties.

Term	Definition
Clod	It is used for a coherent mass of soil broken into any shape by artificial means such as by tillage.
Cohesive bonding	The attraction of similar or like molecules for each other is cohesive bonding.
Complementary ion effect	Is defined as the influence of one adsorbed ion on the release of another from the surface of a colloid.
Complete fertilizer	A fertilizer material which contains all three major nutrients, N, P and K.
Concentrated organic manures	A material of organic origin derived from raw materials of animal or plant, without bulky in nature having no definite composition of plant nutrients.
Concretion	It is coherent mass formed within the soil by the precipitation of certain chemicals dissolved in percolating waters.
Consistency	Consistency is the behaviour of soil under stress. Soil consistency is defined as the manifestation of the physical forces of cohesion and adhesion acting within the soil at various moisture constants.
Consumptive use	The quantity of water lost by evapo-transpiration (ET) plus the contained in plant tissue.
Contact exchange	is defined as the phenomenon by which the exchange takes place between ions held on the surfaces of solid phase particles (solid-solid exchange) and that it does have occur via the liquid phase.
Crumb pores	There are two types of pores involved in the soil aeration-pores between the crumbs called *Intercrumb pores* and pores within the crumbs called *Crumb pores*.
Decomposition	Physical or mechanical weathering process is designated as disintegration and chemical weathering process is designated as decomposition.
Deficient	When an essential element is at a low concentration that severely limits yield and produces more or less distinct deficiency symptoms. Extreme deficiencies will lead to death of the plant.
Denitrification	The microbial reduction of nitrate and nitrite with the release and loss of molecular nitrogen, in some instances of NO_2.

Term	Definition
Diffusion	Diffusion is the molecular transfer of gases.
Dispersed phase	The substance in solution is called dispersed phase.
Dispersion	When the dilute colloidal particles (suspended in water) tend to repel each other, allowing each particle to act completely independent of the others. This is known as dispersion.
Dispersion medium	The medium in which the particles are dispersed is called the dispersion medium.
Edaphology	Study of soil in relation to growth, nutrition and yield of crops.
Eluviation	Is he mobilization and translocation of certain constituents namely, clay, Fe_3O_3, Al_2O_3, SiO_2, humus, $CaCO_3$, and other soluble salts etc. from one area of the soil body to the other area.
Enzyme	An enzyme is a substance, composed of protein, which is capable of lowering the activation energy of other selective compounds enough to allow the breaking of a particular bond under a particular environment. So such reactions influenced by enzymes are called biological reactions.
Excessive	When the concentration of an essential plant nutrient is sufficiently high to result in a corresponding shortage of another nutrient.
Exchange acidity	May be defined as the acidity develops due to adsorbed hydrogen (H^+) and (Al^{3+}) ions on the soil colloids. The magnitude of this exchange acidity is very high.
Farm yard manure (FYM)	Refers to the well decomposed mixture of dung, urine, farm litter (bedding material) and left over or used up materials from roughages or fodder fed to the cattle.
Ferrolysis	Is the process of clay decomposition and transformation under the influence of periodic reduction of iron oxides to ferrous ion (Fe^{2+}).
Fertilizers	Materials having definite chemical composition with a higher analytical value and capable of supplying plant nutrients in available forms.

Term	Definition
Field capacity	The capacity of soil to retain moisture against the downward pull of the force of gravity and moisture is held with soil water potential less than $-1/3$ bar.
Fixation	Fixation of plant nutrients in soil is defined as the process whereby readily soluble plant nutrients are changed to less soluble forms by reaction with inorganic or organic compound of the soil restricting their mobility in the soil and thereby suffer a decrease in their availability to the plant.
Flocculation	The colloidal particles are coagulated by adding an oppositely charged ion. This process of formation of flocks is known as flocculation.
Fragment	It is broken ped.
Gamma (γ) radiation	Are not charged particles, and are not deflected in magnetic and electric fields. It is similar to X- rays but is of shorter wavelength.
Geographic Information system (GIS)	GIS is a system of capturing, storing, storing, checking, integrating, manipulating, analyzing and displaying data which are spatially referenced to the Earth.
Gravitational water	May be defined as the water that is held at a potential greater than $-1/3$ bar and that portion of water that will drain freely from the soil by the force of gravity.
Green manuring	Can be defined as practices of ploughing or turning into the soil undecomposed green plant tissues or the purpose of improving soil physical, chemical and biological environment.
Green manures	May be defined as materials which are undecomposed green plant tissues susceptible to decomposition in the soil after incorporation.
Green manuring in situ	A system by which green manure crops are grown and incorporated into the soil of the same field that is to be green manured, either as a pure crop or an intercrop with the main crop.
Growth	Is defined as the progressive development of an organism.
Gully erosion	Is the removal of soil by running water, with the formation of channels that cannot be smoothed out completely by normal agricultural operation or cultivation?

Term	Definition
Half life period	The time required for a given amount of a radio element to decay to one-half its initial value is called it's half life
Heat capacity	The heat capacity of given material is equal to its specific heat multiplied by its mass.
Heat of solution	When ions are hydrated, a large amount of energy is released and this is known as heat of solution.
Hemicellulose	Are water insoluble polysaccharides.
Humification	Is the process of decomposition of organic matter and synthesis of new organic substances?
Humus	After the active decomposition, organic residues are collectively called humus (organic soil colloid).
Hydration	Chemical addition of water molecules.
Hydrolysis	Involves the splitting of water into H^- ions and OH^- ions.
Hygroscopic water	The water that is held by the soul particles at a suction of more than -31 meters.
Igneous Rocks	They are formed by solidification of molten material magma on or beneath the surface of earth.
Illuviation	Is the immobilization and accumulation of the eluviated constituents at a depth below the soil surface?
Immobilization	Nitrogen immobilization means the change of mineral nitrogen to organic forms by soil micro-organisms.
Incomplete fertilizer	A fertilizer material, which lacks any one of three major nutrient elements.
Insufficient	When the level of an essential plant nutrient is below that required for optimum yields or when there is an imbalance with another nutrient. Symptoms of this condition are rarely observed.
Interaction	The two elements combine to produce an added effect not due to one of them alone in relation to plant growth.
Ion exchange	Is defined as a reversible process by which cations and anion are exchange between solid and liquid phases, and between solid phases it in close contact with each other.

Term	Definition
Isobars	Are atoms of different elements with the same mass number but different atomic numbers? *e.g.*, $_{18}Ar^{40}$, $_{19}K^{40}$.
Isomorphous substitution	Is the substitution of one ion for another of similar size but lower positive valence?
Isotones	Are atoms of different elements having the same number of neutrons but different mass numbers?
Isotopes	Are atoms of the same element whose nuclei contain the same number of protons, but different numbers of neutrons, *i.e.*, the isotopes of the same element having the same atomic number but they differ in mass numbers. *e.g.*, $_8O^{16}$, $_8O^{17}$, $_8O^{18}$, $_{18}Ar^{36}$, $_{18}Ar^{38}$, $_{18}Ar^{40}$, H^1, H^2, H^3.
Leaching requirement (LR)	May be defined as the fraction of the irrigation water that must be leached through the root zone to control soil salinity at any specified level.
Lime requirement	Of an acid soil may be defined as the amount of liming material that is added to raise the p^H to some prescribed value. This value is usually in the range of p^H 6.0 to 7.0.
Liming Factor	May be defined as the factor by which the actual amount of lime can be calculated from the estimated theoretical amount of lime.
Manures	May be defined as materials which are organic in origin, bulky and concentrated in nature and capable of supplying plant nutrients and improving soil physical environment having no definite chemical composition with low analytical value produced from animal, plant and other organic wastes and by products.
Mass flow	Mass flow of air is apparently due to pressure differences between the atmosphere and the soil air.
Maximum water holding capacity	Is defined as the capacity of soil to retain water is exceeded.
Metabolic processes	The mechanisms by which nutrients are converted to cellular material or used for energetic purposes are known as metabolic processes.
Metamorphic rocks	Those rocks which have undergone some chemical or physical change from its original form.

Term	Definition
Milliequivalent	May be defined as one milligram of hydrogen or the amount of any other ion that will combine with or displace it.
Mineralisation	The process by which nitrogen in organic compounds becomes converted into the inorganic ammonium (NH_4^+) and nitrate (NO_3^-) ions carried out by different soils micro organisms. R-NH_2 NH_4 NO_2^- NO_3^- Organic N (Amine) (Ammonium) (Nitrite) (Nitrate).
Minerals	Are naturally occurring inorganic solid homogeneous substances composed of atoms having an orderly and regular arrangement with definite chemical composition and a characteristic geometric form.
Mixed or complex fertilizer	A fertilizer material which contains more than one primary or major nutrient elements produced by the process of chemical reactions.
Moisture equivalent	The percentage of water held by 1 centimeter thick moist layer of soil after subjected to a centrifugal force of 1,000 times gravity for half an hour.
Nitrification	In this process, the so formed ammonia or ammonium salts are converted to nitrate (NO_3^-) form of inorganic nitrogen through intermediate nitrite (NO_2^-) formation as follows: enzymic $2NH_4^+ + 3O_2$ $2NO_2^- + 2H_2O$ $+4H^+$ + Energy oxidation (nitrosomonas) enzymic $2NO_2^- + O_2$ $2NO_3^-$ + energy oxidation (nitrobacter)
Nitrogen	The fertilizer preparation with Rhizobium culture is known as "Nitrogen".
Nutrients	May be defined as the chemical compound required by an organism.
Oxidation	Is the chemical combining of oxygen with a compound and the loss of electrons?
Paleo humus	Is a buried organic substance found in buried soils (paleo sols) that have been derived from remnants of plant and animal life as well as from the products of humification?
Particle density	The weight per unit volume of the solid portion of soil is called particle density.

Term	Definition
Passive transport	Transport of ions from the epidermis to the xylem without any consumption of metabolic energy merely in consequence of the spontaneous disappearance of non-equilibrium driven by a decrease in free enthalpy is known as passive transport.
Peds	Natural aggregates are called peds.
Petrogenesis	Study the origin of rocks.
p^F	May be defined as the logarithm of centimeter height of a water column to give the necessary suction.
Physical weathering	May be defined as he process by which disruption of consolidated massive rocks into smaller bits was found without any corresponding chemical change or formation of new products.
Plant nutrition	Is defined as the supply and absorption of chemical compound required for plant growth and metabolism OR the process of absorption and utilization of essential elements for plant growth and reproduction.
Plasticity	Plasticity is the ability to change shape continuously under the influence of an applied stress and retain the impressed shape on removal of the stress.
Podzolisation	Is a process of soil formation resulting in the formation of podzols and podzolic soils? Podzolisation is the negative of calcification, whereas calcification, tends to concentrate calcium in the lower part of the 'B' horizon. Podzolisation leaches the entire solum.
Porosity	Pore spaces (also called voids) in soils consist of that portion of the soil volume not occupied by solids, either mineral or organic.
Priming action	Is defined as the loss of native soil organic matter through the application of fresh organic residues in the soil.
Protein	The protein molecule is composed of a long chain of amino acids having general structured H_2N CHRCOOH where R may be a hydrogen atom, a single methyl group, a short carbon chain or a cyclic structure.

Term	Definition
Radioactivity	Is a phenomenon in which nuclei of certain elements undergo spontaneous disintegration with emission of an a and b particles and the formation or synthesis of the nucleus of a new element.
Reduction	Is the chemical process in which electrons are gained, the negative charge is increased and the positive charge is decreased.
Remote sensing	Is defined as the measurement or acquisition of information of some property of an object or phenomenon, by a recording device that is not in physical or intimate contact with the object or phenomenon under study.
Rill erosion	Is the removal of surface soil by running water, with the formation of narrow shallow channels that can be leveled or smoothed out completely by normal cultivation?
Rocks	The mixtures of two or more minerals.
Saltation	Is a process of soil movement in a series of bounces or jumps?
Saturated flow	When soil water moves mainly due to gravity, which is at moisture potentials greater than $-1/3$ bar, the movement is called saturated flow.
Sedimentary Rocks	They are formed from consolidation of sediments derived from the breaking down of pre-existing rocks.
Sheet erosion	Is the removal of a fairly uniform layer of surface soil by the action of rainfall or runoff water?
Soil	Soil is a natural body developed by natural forces acting on natural materials. It is usually differentiated into horizons from mineral and organic constituents of variable depth, which differ from the parent material below in morphology, physical properties and constituents, chemical properties and composition and biological characteristics.
Soil acidity	May be defined as the soil system's proton (H^+ ions) donating capacity during its transition from a given state to a reference state.
Soil aeration	Exchange of carbon dioxide and oxygen gases between the soil pore space and the aerial atmosphere.

Term	Definition
Soil Biology	The study of effect of plants, animals and soil micro-organisms on the evolution, chemical composition and physical condition of the soil.
Soil chemistry	The study of chemical components of soil, their interaction with one another and the effects of the chemical environment of soil.
Soil colloids	soil particles less than 0.001 mm size possess colloidal properties and are known as soil colloids.
Soil crusting	Is a phenomenon associated with deterioration of soil structure, where the natural soil aggregates break and disperse?
Soil erosion	Is defined as the detachment and transportation of soil mass from one place to another through the action of wind, water in motion or by he bating action of rain drops.
Soil formation	Soil formation is a process of two distinct phases: weathering of rocks and mineral, *i.e.,* disintegration (physical) and decomposition (chemical) of rocks and minerals the development or the formation of true soil by some soil forming factors and pedogenic processes.
Soil Geology	The study of geological materials from which the soils is derived and it' process of formation.
Soil Physics	Study the effects of physical laws on the evolution of chemical properties and physical properties of soil.
Soil profile	A vertical section through a soil, represents sequence of horizons or layers differentiated from one another but genetically related and included to the aren't material from which soil profile developed.
	Science of rocks.
	Description of rocks.
Soil reaction	Is an indication of the acidity or basicity of the soil and it has been classified into three groups: (i) Acidity (ii) Alkalinity and (iii) Neutrality.
Soil separates	The various groups like gravels, sands, silts, clays are termed as soil separates or fractions.
Soil structure	The arrangement of primary particles and their aggregates into a certain definite pattern.

Term	Definition
Soil survey	Is a definite study of soil morphology in the field, corroboration of diagnostic soil properties in the laboratory, classification of soils of the area in well defined units, plotting their extent and boundaries on a map, and prediction of the adaptability of these soils to various uses
Soil texture	Refers to the relative percentage of sand, silt and clay in a soil.
Specific heat	May be defined as the amount of heat required to raise the temperature of 1 gram of substance by 1°C.
Splash erosion	Soil splash caused by the impact of falling rain drops.
Starch	Is the polymer of glucose and it serves the plant as a storage product, and as such it is the major reserve carbohydrate, plant starches usually contain two components, amylose and amylopectin.
Straight fertilizer	Chemical fertilizers, which contain only one primary or major nutrient element.
Stream channel erosion	Is the scouring of material from the water channel and the cutting of banks by flowing or running water?
Submerged soils	Are soils that saturated with water for a sufficiently long time in a year to give the soil the following distinctive gluey horizons resulting from oxidation – reduction processes: A partially oxidized 'A' horizon high in organic matter. (i) a mottled zone in which oxidation and reduction alternate and (ii) A permanently reduced zone which is bluish green in colour.
Suction	The force with which water is held is termed as suction.
Superfluous water	The water which is retained in the soil beyond the field capacity soil moisture tension.
Surface creep	Is the rolling or sliding of large soil particles along the ground surface?
Surface tension	Is generally evidenced at water air interfaces and it may be defined as the forces in dynes acting at right angles to any line of 1 cm. Length in the surface. At the surface, the attraction of air for the water molecules is much less than that of water molecules for each other. Consequently there is a net downward (in ward) force on the surface molecules, resulting in sort of a compressed film at the surface. This phenomenon is called surface tension.

Term	Definition
Suspension	Represents the floating of small sized particles in the air stream.
Threshold velocity	Is defined as the minimum velocity required initiate movement from the impact of soil particles carried in saltation.
Total acidity	Summation of active and exchange acidity.
Toxic	When the concentration of either essential or other elements is sufficiently high to inhibit plant growth to a great extent. Severe toxicity will result in death of plant.
Unavailable water	The water which is held at soil water potential greater than -15 bars.
Unsaturated flow	It is the low of water held with water potential lower than $-1/3$ bar. Water will move toward the region of lower potential.
Water repellant soils	When fatty or oily substances, which are low in oxygen, coat the soil particles, water is not attracted to and held to the coated surface such soils are called water repellant soils.
Water use efficiency (WUE)	The amount of water required to produce a unit of dry weight material, *e.g.*, kilogram corn, is a measure of water use efficiency.
Water vapour movement	**Internal movement:** The change from the liquid to the vapour state takes place within the soil that is in the soil pores. **External movement:** The phenomenon occurs at the land surface and the resulting vapour is lost to the atmosphere by diffusion and convection.
Weathering	Is an inevitable natural process of breakdown and transformation of rocks and minerals into unconsolidated residues (regolith), lying on the surface of the earth, with varying depth.
Wilting coefficient	That amount of water which is held with water potential less than -15 bars and it is held so strongly that plants are not able to absorb it for their needs.
Zoung soil	The soil where the soil forming factors and pedogenic processes are still in operative conditions and changing the properties of soil in the profile and the processes have not made a prominent impression on the soil profile.

4

REASONING :
PRINCIPLES AND CONCEPTS

(1) Stokes's Law

G.G. Stokes's (1851) suggested the relation between the radius of a particle and its rate of fall in a liquid. He stated that the resistance offered by the liquid to the fall of the particle varied with the radius of the sphere and not with the surface.

According to formula, the velocity of fall of a particle with the same density is directly proportional to the square of the radius and to the square of the radius and inversely proportional to the viscosity of the medium.

$$V = 2/g \ (dp - d) \ gr^2/n$$

Where,

V = velocity of fall (cm/sec^2)

G = acceleration due to gravity (cm/sec^2)

dp = density of the particle (g/cc)

d = density of the liquid (g/cc)

r = radius of the particle (cm)

n = absolute viscosity of the liquid (poise or m poise)

DERIVATION OF STOKE'S LAW

If a solid body (soil particle is moved through a liquid, the thin layer of liquid in immediate contact with the solid is virtually at rest, just as in the flow of liquid through a tube; as a result of viscosity a viscous drag is exerted on the moving body. In order to maintain a uniform velocity a steady force must, therefore be applied to overcome the influence of the viscosity of the liquid.

It has been found that if a small sphere or soil particle of radius 'r' travel at a velocity 'v' through a fluid, gas or liquid, having a coefficient of viscosity 'n', the force applied 'f', which just balances that due to viscosity given by Stokes's law,

$$f = \sigma \pi r \, n \, v \tag{1}$$

If the sphere or soil particle is falling under the influence of gravity force, the constant downward force is,

$$4/3 \times \pi r^3 \, (dp - d) \, g \tag{2}$$

according to the opposing force of viscosity increases with increasing rate of fall of solid body or soil article and eventually a constant speed will be attained when the viscous force (1) is exactly equal to the gravitational pull (2), that is

$$4/3 \, \pi r^3 \, (dp - d) \, g = \sigma \pi r \, n \, v$$

$$V = \frac{2}{g} \frac{(dp - d) g r^2}{n}$$

This form of Stoke's law is applicable to a solid sphere or soil particle falling through a liquid or gas or to a drop of liquid falling through a gaseous medium.

ASSUMPTIONS IN STOKE'S LAW

The particle must be large in comparison to liquid molecules so that Brownian movement will not affect the fall.

The extent of the liquid must be great in comparison with the size of the particles.

Particles must be rigid and smooth.

There must be no slipping between the particle and the liquid.

The velocity of fall must not exceed a certain critical value so that the viscosity of the liquid remains the only resistance to the fall of the particle.

Particles greater than silt size fractions of a soil mass cannot be separated accurately with the help of this Stoke's law.

LIMITATIONS OF STOKE'S LAW

The effect of different particle shapes on the settling velocities of clay particles is a major limitation for the accuracy of this law.

During mechanical analysis based on this principle, it is necessary to maintain a known constant temperature because the rate of tall varies inversely with the viscosity of the medium, which changes with the change in temperature.

The density of the soil particle is another factor that affects the accuracy of this law. Density depends upon the mineralogical and chemical constitution of the particles and their degree of hydration.

(2) Fick's Law

According to Fick's law, diffusion is a function of the concentration gradient, the diffusion co-efficient of the medium, and the cross-sectional area. D

$$dQ = DA \left(\frac{dc}{dx}\right) dt$$

where, dQ is the mass flow (moles) during the time at across area A (sq. cm.), dc/dx the concentration gradient [moles/c.c. (cm)], and D, the proportionately constant or diffusion coefficient (sq. cm./sec). 'D' depends upon the property of the medium as well as the gas. It varies directly with the square of the absolute temperature and inversely with the total pressure.

(3) Laws of plastic flow

The essential difference between viscous and plastic flow is that a certain amount of stress must be added of plastic soils before flow is produced.

The following equation shows that the volume flow is a function of the force applied.

$$V = K\mu(F - f)$$

where,

V	=	volume of flow
μ	=	co-efficient of mobility.
F	=	applied force.
f	=	force necessary to overcome the cohesive forces of the system and just enough to start the flow (this force is termed "yield value")
K	=	constant

The flow of viscous nature when the value of 'f' is O and then the volume of flow is proportional to the applied force and the co-efficient of viscosity of the liquid. Curve AB show the viscous flow and it increases directly with applied pressure. Curve APQR show a plastic flow and here a certain force is applied to start the flow and then force is proportional to the applied force (as shown in segment QR). The yield value is obtained by extrapolating the segment QR to the point S on the abscissa. The magnitude of yield value is correlated with the extent of the cohesive forces of the water films between the particles.

(4) Jarusov's Rule

The exchangabily of given ion was not a constant, but was a function of its amount in relation to the amount and kind of other ions present.

In terms of bonding energy the consequences is that where two different cations are present, the cation having the higher bonding energy pre-empts those positions on the surface or that part of the diffuse layer which correspond to the greatest energy release, leaving for he cation of the lesser bonding energy only those positions which yield smaller release. In this way one ion influences the activity of another through the fee energy relationships of the diffuse double layer and the ionizing surface.

(5) Plank's Law

According to Planck's law, all bodies a temperature below 0^0 absolute emit electromagnetic radiation at various wavelengths,

$$Wr = \frac{8\bar{x}\,hc}{\lambda^5} \frac{1}{\left[C^{ch/\lambda kt} - 1\right]}$$

where,

Wλ = energy radiated from a black body of wavelength λ

h = height/latitude

c = velocity of light

T = absolute temperature

K = gas constant

The earth can be treated as a blackbody at ~300 k emitting electromagnetic radiations with peak emission at about 9.7μm. According to Planck's law, the radiation emitted by the earth (300 k) is much less at all wavelengths as compared to emitted radiation from sun (6000 k). However at the earth's surface because of the great distance between the sun and the earth, the energy in between 7-15 μm wavelength region is predominant due to the thermal emission of the earth.

(6) Liebig's law of minimum

In 1862 Justas Von Liebig, a German chemist states that "Energy field contains a maximum of one or more and a minimum of one or more nutrients with this minimum, be it time, potash, nitrogen, phosphoric acid, magnesia or any other nutrient, the yields stand in direct relation. It is the factor that governs and controls yields. Should this minimum be time——yield——will remain the same and be no longer even though the amount of potash, silica, phosphoric acid etc. be increased a hundred fold."

This law can be simply stated as follows:

"Even if all but one of the essential elements be present, the absence of that one constituent will render the crop barren".

(7) Mitscherlich's Law

E.A. Mitscherlich (1909) was among the first to quantify the relationship between plant growth response and the addition of a growth factor. He developed two laws from his research works which are as follows:

(i) *Law of physiological Relationships*

"Yield can be increased by each single factor even when it is not present in the minimum as long as it is not present in the optimum".

(ii) *Growth Law*

"Increase in yield of a crop as a result of increasing in a single growth factor is proportional to the decrement from the maximum yield obtainable by increasing the particular growth factor."

(8) Darcy's Law

Darcy stated that the rate of flow increased with an increased depth of water above the bottom of the soil and decreased with an increased depth of soil through which water flowed and can be expressed as follows:

$$Qw = -k\frac{(\Delta dw)\,At}{(\Delta ds)}$$

where,

Qw = quantity of water in C.C.

K = hydraulic conductivity in cm sec^{-1}.

Δdw = water or hydraulic head in cm.

A = soil area in sq. cm.

t = time in seconds

Δds = soil depth in cm.

(1) Concept of Soil

The word 'soil' as a verb means 'to make dirty' as in the case of soiled dishes or clothing. The noun soil is derived through old French from the Latin 'solum', which means floor or ground. What a soil scientist called soil, a geologist may call fragmented rocks, an engineer may call earth and an economist may call land.

There are two basic concept of soil that has already evolved through two centuries of scientific study. The first one considered soil as a natural body, a biochemically weathered and synthesized product of nature and the second one considers soil as a natural habitat for plants and other living organisms and justifies soil studies primarily on that basis. So, there are two approaches in studying soils one by pedologist which includes the study of origin of the soil, its classification and its description and the other by pedapologist which covers the study of soil in relation to growth, nutrition and yield of crops.

(2) Principle of mechanical analysis

Soil consists of particle of various sizes and since the fundamental objectives of a particle size analysis is to determine the percentage distribution of those particles (sand, silt and clays) in the soil mass. The rate of fall of particles in a viscous medium depends upon the size, density and shape of the particle. In a medium like water, larger particles settle more rapidly as compared to smaller ones with the same density and consequently settle out of suspension very quickly. This principle serves the basis of practically all mechanical analysis.

(3) Soil water Energy Concepts

The retention and movement of water in soils, its uptake and translocation in plants and potential evapotranspiration etc. are also related to energy. Different kinds of energy are involved including potential, kinetic and electrical. By using the term 'free energy' (ability to do work) energy status of water can be characterized to indicate the strength with which water is held.

Recently 'soil water potential' is used to measure the pressure required to force the water off soil and it may be defined as the work the water can do when it moves from its present state to a pool of water in the reference state. The movement of water in soil takes place from a higher free energy to a lower free energy level.

The soil water potential is a combination of the effects of the surface area of soil particles and small soil pores that adsorb water, metric potential (ψm) the effects of attraction of ion and other solutes for water, solute or osmotic potential (ψs) and the atmospheric ort gas pressure effects, pressure potential (ψp). In salt free well drained soil, metric potential is almost equal to the soil water potential (ψw). An additional effect of the position of the water (such as being elevated) compared to the reference state (the reference free energy state = o and is at a specified elevation) is called the gravitational potential (ψg). Gravitational potential is not related to soil properties, only to the elevation of water in comparison to a reference poison. Various potentials can be written as follows:

ψw	=	ψm	+	ψs	+	ψp
Soil water potential		metric potential		solute or osmotic potential		pressure potential

$$\psi t \quad = \quad \psi w \quad + \quad \psi g$$

total water potential	soil water potential	soil gravitational potential

Most o the productive soils have no depth of water standing on it and can be written as follows:

$$\psi t \quad \approx \quad \psi w \quad \approx \quad \psi m$$

Therefore, among all potentials metric potential (ψm) is the most important and dominant for more soils.

(4) Film theories of plasticity

The colloidal clay particles in the soil act as a lubricant between coarser particles and diminish their friction. It is highly probable that the plate shaped particles are oriented in such a way that their flat surfaces are in contract. This orientation increases the amount of contact between the colloidal particles. The increased contact together with the raising of the ratio of water film surface to the particle mass, may be considered as producing the plastic effects within a certain moisture range, the tension effects of the water films between the oriented plate like colloidal particles, which impart to the soil its cohesion phenomenon, enable the soil to be molded into any deformed shape. This moisture range corresponds to the range of plasticity of a soil. Orientation of particles and their subsequent sliding over each of the takes place when sufficient water has been added to provide a film around each particle. The amount of water required to produce these films corresponds to the moisture content at which the soil ceases to be friable. With an excess of water, the water films becomes so thick that the cohesion between particles decreases and the soil mass becomes viscous and flows.

(5) Soil water Energy Concepts

(I) Size and chemical composition

The chemical analysis of clay indicates the presence of silica, alumina, iron and combined water. These make up from 90-98 per cent of the colloidal clay. The soil colloidal matter contains plant nutrients like Ca, Mg and K etc. clay is a mixture of hydrated aluminoferro silicates of varying composition mixed in some cases with the excess of sesquioxides or silica.

The term 'clay' has three meaning in soil usage:

It is particle fraction composed of any particles fraction composed of any particles less than 2 microns ($<2\mu$) in effective diameter.

It is a name for minerals of specific composition;

It is a soil textural class.

(ii) Shape

Silicate clay minerals have been examined by electron microscope and found that the particles are laminated made up of layers of plates or flakes or even rods. Each clay particle is made up of a large number of plates like structural units. The different units or flakes of clay minerals are held together with varying degrees of force depending upon the nature of the clay minerals. The edges of some clay particles are clean cut and others are frayed or fluffy. In all cases, clay minerals are developed more in the horizontal axis than that of vertical axis.

(iii) Surface area

The surface area of a clay particle is usually defined as the area of the particle that is accessible to ions ore molecules when the clay is an aqueous solution.

All clay particles must expose a large amount of external surface. In some clay there are extensive internal surfaces as well. This internal exists between the plate like crystal units that make up each particle so the large surface area of clay colloids is not only due to its fineness but also its plate like structure.

(iv) Electronegative charge

Clay micelles (micro cells) carry negative charges and so a number of oppositely charged ions (cations) are attracted to each colloidal clay crystal. The colloidal clay particles have inner ionic layer (surface of highly negative charge) and the outer ionic layer (cations swarming layer).

(v) Adsorbed cations

Clay micelles adsorb a number of cations-humid, arid and semi-arid regions colloids- cations are H^+, Al^{3+}, Ca^{2+}, Mg^{2+} Na^+ and K^+ the amount of these cations held by clay vary with its kind. Cations adsorbed (if dominant on the clay colloids very often determines the physical and chemical properties of the soil and thereby influence the plant growth).

(6) Milliequivalent concept

For the measurement of Cation Exchange Capacity (CEC), the term equivalent or more especially "Milliequivalent" is used because the number of negative charge sites in a given soil sample does not change, but the weights of the cations that may be adsorbed to those sites at one time do change because they have different weights.

The term 'milliequivalent' may be defined as one milligram of hydrogen or the amount of any other ion that will combine with or displace it. The milliequivalent weight of a substance is one thousand[th] of its atomic weight. Thus, if a clay has a cation exchange capacity of 1 milliequivalent (1 me/100 g), it is capable of exchanging 1 mg of hydrogen or its equivalent for every 100 g of clay. The weight of one hectare furrow slice (depth 0-15 cm) is 2.2×10^6 kilogram. Therefore, one hectare furrow slice (depth 0-15 cm) soil is capable of exchanging 22 kilograms of hydrogen.

Calculation:

0.1 kg soil can exchange $1/10^6$ kg of hydrogen.

Kg soil can exchange $1/0.1 \times 10^6$ kg of hydrogen.

2.2×10^6 kg soil can exchange $2.2 \times 10^6/ 0.1 \times 10^6$ of hydrogen.

= 22 kilogram of hydrogen

(7) Elementary Concepts of Diffuse Double Layer

Helmholtz	Gouy	Stern
+ + + + + + +	+ + + + + + +	- + - + - + +-
- - - - - - -	+ + + + + + +	+ - + - - + --
- - - - - - -	+ + + + + + +	+ - + + - + +-
	+ + + + + + +	- - - - - -
- - - - - -	+ + + + + + +	+ - + + - + +-
	+ + + + + + +	- - - - - -
+ + + + + + +	+ + + + + + +	+ - + - + - +
O → x	o　　→　　x	o　　→　　x

Distance from Particle Surface

The exchangeable ions are surrounded by water molecules and may thus be considered as forming a solution which is often a micellar solution or inner solution. The solution containing free electrolytes called outer solution or intermicellar solution. The ionic conditions on the outside surface of a clay particle or pocket of clay particles dispersed in water or an electrolytes solution are controlled by proportion of the exchangeable cations that disperse into the solution. A clay surface probably behaves as an effectively unchanged surface if the negative charge on the lattice is neutralized by monovalent cations that are tightly bound to the surface. But if the cations are hydrated a proportion tends to dissociate from the surface and will cause on electrical potential gradient to be set up near the surface. The system clay lattice exchangeable cations solution can be looked upon as forming a complex electrical double layer, known as the 'helmholtz double layer', the inner layer being the surface of the lattice carrying the negative charge and the outer layer being composed of two parts a positive layer due to the cations bound to the lattice surface known as the fixed layer or 'stern layer' and a positive layer diffused in the solution close to the lattice surface is the 'Gouy diffuse layer.'

(8) Principles of Remote Sensing

There are various stages of remote sensing which are also involved in the principle of remote sensing. Some of those are given below.

(i) Origins of electromagnetic energy (*e.g.*, sun, transmitter carried by the sensor).

(ii) Transmission of energy from the source of the surface of the earth and its subsequent interaction with intervening atmosphere.

(iii) Interaction between energy and earth surface or self-emission.

(iv) Transmission of the emitted or reflected energy to the remote sensor.

(v) Detention of the energy by the sensor converting into photographic image or electrical output.

(vi) Recording of the sensor output.

(vii) Data processing for the generation of date base.

(viii) Collection of ground truth and other collateral information.

(ix) Date processing and interpretation.

(9) Principles of liming reactions

Lime reactions in soils depend upon the nature and the fineness of the liming materials. Lime is usually applied to soils in the form of ground limestone. Limestone can be classified as calcite ($CaCO_3$), dolomite [$CaMg (CO_3)_2$] or a mixture of the two. Both of these limestones are sparingly soluble in pure water but do become soluble in water containing carbon dioxide. The greater the partial pressure of carbon dioxide in the system, the more soluble the limestone becomes. Dolomite is somewhat less soluble than calcite. The reaction of limestone ($CaCO_3$) can be written as

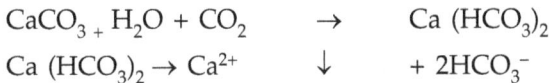

$$CaCO_{3+} H_2O + CO_2 \rightarrow Ca (HCO_3)_2$$
$$Ca (HCO_3)_2 \rightarrow Ca^{2+} \quad \downarrow \quad + 2HCO_3^-$$

(takes part in cation exchange reactions)

In this way hydrogen ions (H^+) in the soil solution react to form weakly dissociated water, and the calcium (Ca^{2+}) ion from limestone is left to undergo cation exchange reaction. The acidity of soil is, therefore, neutralized and per cent base saturation of the colloidal material is increased.

(10) Salt precipitation Theory

Recently salt precipitation theory is employed satisfactorily for the reclamation of sodic soils. The elimination of salts and exchangeable sodium from soils by leaching is presently practicing, but the leached salts have been washed into ground waters or streams, making those waters more salty and again that too much salty water is used for irrigation purpose. Due to such use the soils are further subjected to salt problems. With this view, a new concept in managing salty soils has been developed and that is known as 'precipitation of salts.'

(11) Flaig's concept of humus formation

Lignin, freed of its linkage with cellulose during decomposition of plant residues, is subjected to oxidative splitting with formation of primary structure units (derivatives of phenyl propane)

The side-chains of the lignin-building units are oxidized, demethylation occurs, and the resulting polyphenols are converted to quinines by polyphenol oxidase enzymes.

Quinines arising from lignin (as well as from other sources) react with N-containing compounds to form dark-coloured polymers.

(12) Kononova's concept of humus formation

There are three stages leading to the formation of humic substances which are as follows:

Stage 1: Fungi attack simple carbohydrates and parts of the protein and cellulose in the modularly rays, cortex of plant residues.

Stage 2: Cellulose of the xylem is decomposed by aerobic myxobacteria. Polyphenols synthesized by the myxobacteria are oxidized to quoins by polyphenoloxidase enzymes, and the quinines subsequently react with n compounds to form brown humic substances.

Stage 3: Lignin is decomposed; phenols released during decay also serve as source materials for humus synthesis.

(13) Bray's Nutrient Mobility Concept

This concept has been proposed by R. Bray and his associates after modifying the concept developed by Mitscherlich-Baule-Spillman. Bray's concept states that "as the mobility of a nutrient in the soil decreases, the amount of that nutrient needed in the soil to produce a maximum yield (the soil nutrient requirement) increases from a variable net value, determined principally by the magnitude of the yield and the optimum percentage composition of the crop, to an amount whose value tends to be a constant."

The magnitude of this constant is independent of the amount of crop yield, provided that the kind of plant, planting pattern and rate, and fertility pattern remain constant and that similar soil an seasonal conditions prevail bray has modified the Mitscherlich equation to:

$$\text{Log } (A - Y) = \log A - C_1 B - C_X$$

Where, A, Y and X are maximum yield, yield obtained and amount of added fertilizer nutrient respectively,

b = amount of an immoblise but available form of nutrients (like P and K)

C_1 = constant or efficiency factor of 'b' for yields.

C = constant or efficiency factor of x.

(14) P^E Concept

It is more logical and convenient to use p^E instead of Eh in the study of redox equilibira. The common reagent in redox equilibria is the electron. Just as p^H is a measure of proton (H^+) activity, so in P^E, the negative logarithm of the electron activity, a measure of electron activity. It can be written as:

$$P^E = -\log (e)$$

$$= Eh/2.303 \text{ RT F}^{-1}$$

$$P^E = Eh/0.0591 \text{ and } P^{Eo} = E_o/0.0591 \text{ at } 25°C$$

So, for the determination of P^E value of redox potential (Eh) divided by 0.0591 will give the P^E value.

(15) Soil Test Crop Response and Targeted Yield Concept

Soil testing is one of the accepted methods for the economic use of fertilize but these are many problems in making fertilizer recommendations based only on soil test values.

Recently in our country systems of soil test rating is being modified incorporating crop response data available from systematic field experiments. It has been accepted by the scientist that any soil test method intended for use in advisory work needs to be correlated with actual crop response obtained under field conditions, and the success of the fertilizer recommendations programme will depend on the accuracy of the calibrations obtained this way. Modern approaches of soil fertility evaluation are mainly focused towards increasing fertilizer use efficiency. The approaches may be as follows:

Soil analysis and correlation

Critical soil test level approach,

Agronomic approach,

Soil fertility cum soil survey,

Inductive approach based on soil test and crop response correlation,

Deductive approach based on soil test and crop response correlation, and

Targeted yield concept approach.

From the soil test, crop response field experiments, it has been possible to derive three basic parameters like (i) nutrient requirement in kg per quintal of the produce, (ii) percentage contribution from soil available nutrients and (iii) percentage contribution from added fertilizers towards making effective fertilizer prescriptions for specific yields.

The parameters have been calculated as follows:

(i) Nutrient requirement kg for producing one quintal of grain =

Total uptake of nutrient in kg/ha (Grain + straw)/Yield of grain (q/ha)

(ii) Contribution from fertilizer (ct) (in %) = total uptake of nutrient in treated plot – (available soil test value of nutrient in treated plot x CS)

(iii) Contribution percentage from fertilizer = cf/fertilizer dose applied in kg/ ha x 100

(16) Principles of some determinations and instruments

(1) Determination of Available, Ammonical and Nitrate Nitrogen in Soil

Principle: The organic matter in the soil is oxidized by KM_nO_4 in presence of NaOH. The ammonia released during oxidation is absorbed in boric acid to convert the ammonia to ammonium orate. The ammonium borate formed is titrated with standard H_2SO_4 required for reaction with ammonium borate, the n is calculated.

(2) Soil testing for available phosphorus (Olsen and Sommers, 1982)

Principle: Phosphorus is extracted from the soil with 0.5 M $NaHCO_3$ at nearly constant p^H 8.5. In calcareous, alkaline or neutral soils containing calcium phosphates, this extractant decreases the concentration of Ca in solution by causing precipitation of Ca as $CaCO_3$ as a result the reactive (high specific surface) form of P is extracted from the phosphates of iron, aluminium and calcium present in the soil.

The heteropoly complexes are thought to be formed by co-ordination of molybdate ion, with P as the central coordinating atom, the oxygen of the molybdate radical being substituted for that of PO_4.

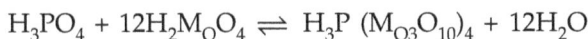

$$H_3PO_4 + 12H_2M_OO_4 \rightleftharpoons H_3P (M_{O3}O_{10})_4 + 12H_2O$$

A characteristic blue colour (the molybdenum blue reaction) is produced when either molybdate or its heteropoly complexes are partially reduced. Some of the molybdenum ions 6^+ are reduced to a lower valence state, probably 3^+ and/or 5^+ valence state involved unpaired electrons from which spectrophotometric response (blue colour) would be expected. Complex forms are reduced by ascorbic acid or stannous chloride. Here ascorbic acid is used as reducing agent.

(3) *Principle of Spectrophotometry*

When light (monochromatic heterogeneous) is incident upon homogeneous medium, a part of the incident light is reflected, a part is absorbed by the medium and the remainder is allowed to transmit as such.

Absorption is generally governed by two separate laws known as Lambert's and Beer's law.

Lambert's Law: When a beam of light is allowed to pass through a transparent medium, the rate of decrease of intensity with the thickness of medium is directly proportional to the intensity of the light.

Beer's Law: The intensity of a beam of monochromatic light decreases exponentially with the increase in concentration of the absorbing substance arithmetically.

(4) *Determination of available potassium from soil Knudsen and Peterson, 1982*

Principle of flame photometer: When a solution of the metallic salt is atomized into a non-luminous flame, elemental K atoms get excited and emit light when come to ground state. The light emitted is filtered through a glass filter which allows light of definite wavelength of the element, 766.5 mm for K, to pass. The light falls on to photocell emitting electrons generating an electric current. This current is measured on galvanometer and is proportional to the concentration of metal element present in the solution atomized.

(5) *Principle of AAS (Atomic Absorption Spectrophotometer)*

The AAS is based on the principle that atoms of metallic elements (Fe, Mn, Zn, and Cu) which normally remain in ground state under flame conditions absorb energy when subjected to radiations of specific wavelength. The absorption of radiation is proportional to the concentration of atoms of that element. The absorption of radiation by the atoms is independent of the wavelength of absorption and temperature of the atoms.

These two features provide AAS distinct advantage over flame emission spectroscopy. It has also greater sensitivity and accuracy.

(6) *Determination of organic carbon from soil*

Wet oxidation methods (Nelson and Sommers, 1982).

Principle: The organic C in organic matter is oxidized by known excess chromic acid ($K_2Cr_2O_7 + H_2SO_4$). The excess chromic acid not reduced by organic matter is determined by back titration with standard $FeSO_4$ solution (redox titration) using ferroin indicator. The organic 'C' in soil is calculated from the chromic acid utilized (reduced) by it.

(7) *Determination of soil pH*

Principle: A glass surface in contact with hydrogen ions of the solution under test acquires an electrical potential, which depends on the concentration of H^+ ions. A measure of the electrical potential (emf) is therefore, H^+ ion concentration or p^H of the solution.

(8) *Determination of soil EC: (Electrical conductivity)*

Principle: A simple Wheatstone bridge circuit is used to measure EC by null method. The bridge consists of two known and fixed resistance r_1, r_2 one variable standard resistance r_4 and the unknown r_3. The variable resistance r_4 is adjusted until a minimum or zero current flows through the AC galvanometer at equilibrium.

$$r_1/r_2 = r_3/r_4$$

Since conductivity is reciprocal of resistive, it is measure with the help of r_3. The electrical conductivity of soil water system rises with increasing content of soluble salts in the soil.

Brief Information Regarding Agrochemicals

Agrochemicals is a broad term which means the chemicals used in agriculture for various purposes viz. plant protection, plant nutrition, plant growth retardation or for plant growth enhancement. A broad group of such chemicals which are used in agriculture falls under the category of pesticides. These pesticides may be of organic or inorganic in origin. From the plant nutrition point of view, a broad category of chemical substances consists of fertilizers. Again, the fertilizers may be of organic or inorganic in origin. Another group of substances of chemical origin used in agriculture constitute mostly of plant growth regulators which either enhance or retard the plant growth; however these substances are also included in the group of pesticides.

PESTICIDES

Any substance or mixture of substances intended for preventing, destroying, or controlling any pest, including vectors of human or animal diseases, unwanted species of plants or animals, causing harm during or otherwise interfering with production, processing, storage, transport or marketing of food, agricultural commodities, wood and wood products or animal feed stuffs.

The term also includes substances intended for use as the plant growth regulator, defoliant, desiccant, fruit thinning agent or an agent for preventing premature fall of fruit and substances applied to crops either before or after harvest to prevent deterioration during storage or transport.

Types of Pesticides

1. **Insecticides :** Substances that prevent destroy or kill insects.

2. **Fungicides :** Substances that prevent destroy or inhibit growth of fungi or diseases in crops.

3. **Herbicides :** Substances used for preventing or inhibiting growth of plants or for killing weeds.

4. **Rodenticides :** Substances that inhibit destroy or kill rodents.

5. **Nematicides :** Substance that prevent, destroy, repel or inhibit nematodes.

6. **Chemo-sterilants :** Substances that sterilize insect-pests, e.g., Diurea, Tepa, Metpa.

7. **Molluscicides :** Substances that prevent, repel, destroy or inhibit growth of member belonging to phylum Mollusca.

8. **Plant growth Regulators :** Substance that cause acceleration or retardation of rate of growth or rate of maturation of plants, e.g., GA_3 Auxins, Cytokine, Abscisic acid, Ethylene.

9. **Defoliants :** Substance that cause plant leaves to die and fall away.

10. **Desiccants :** Substance that cause draining of moisture out of plants, causing them to dry, e.g., Diquat, Paraquat, Cacodylic acid, Dinoseb.

11. **Attractants :** Substances that attract insect pests, e.g., Gyplure, Gossyplure.

12. **Repellents :** Substance that repel, insect pest from a treated plants, e.g., Deet, Dimthyl phthalate.

INSECTICIDES

These are chemical substances which protect plants against invasion of insects or are used to eradicate existing infection of insects.

Classification of insecticides on the basis of chemical nature

Inorganic

A. Arsenicals
   (i)   Lead arsenate
   (ii)   Calcium arsenate
   (iii) Sodium arsenate
   (iv)   Paris green
B. Flurides
   (i)   Sodium fluride
   (ii)   Sodium flurosilicate
   (iii) Sodium fluroaluminate
C. Sulphur
   D.   Lime Sulphur
   E.   Zinc Phosphate

Organic

A. **Hydrocarbon oil** : For example, Kerosene, Petroleum.
B. **Plant originated compounds** : For example, Azadiractin, Rotenone, Rhynodine.
C. **Animal originated compounds** : For example, Nereistoxon.
D. **Organophosphate compounds** : For example, Quinolphos, Monocrotophos.
E. **Carbamate** : For example, Carbaryl, Carbofuron.
F. **Synthetic pyrethroids** : For example, Cypermethrin, Oeltamethrin.
G. **Chlorinated hydrocarbon compound**
   (i)   Organochlorine e.g., DOT.
   (ii)   HCN   e.g., BHC.
   (iii) Cyclodine e.g., Endosulphan, Eldrin.
   H. **Fumigant** e.g. Methyl bromide, CCl_4, Ethylene dibromide.

FUNGICIDES

Fungicides are agent of natural or synthetic origin, which act to protect plants against invasion by fungi and or to eradicate established fungal infection.

Classification of fungicides on the basis of mode of action.

1. **Systemic fungicide :** Is one which is taken up and translocated within plant as a result of which the later becomes fungitoxic e.g., Carbendazim, Benomyl.

2. **Non-systemic :** Non-systemic fungicides do not penetrate the plant. On application they kill pathogen on the surface of foliage and fruits, e.g., Zineb, Thirum.

Classification on the basis of chemical composition.

Systemic

1. **Oxathilin and related compounds :** Carboxamides, Carboxin (Vitavax).

2. **Benzimidazoles :** Carbendazim, Benomyl.

3. **Thiophanates :** Thiophanate methyl.

4. **Morphalines :** Tridemorph

5. **Pyrimidines :** Fenarimol

6. **Triazole Compounds :** Triadimeton, Bitertanal.

7. **Organo phosphorus compounds :** Kitazin.

8. **Piperazine-T :** Triforine.

Non-Systemic

1. **Sulphur fungicides :** Wetasulf, Sulphur dust, Sulked, Thiovit.

2. **Dithiocarbamates :** Zine, Thiram, Ziram.

3. **Copper fungicide :** Bordeaux mixture, copper oxychloride.

4. Mercury fungicides

   Inorganic : Mercuric chloride, Mercurous chloride

   Organic : Methoxy ethyl mercury chloride (Agallol), Ceresol.

5. **Heterocyclic nitrogen compounds :** Captan, Captatol, Glyodin.

HERBICIDES

Chemical compounds used for killing or reducing the growth of weeds are called as herbicides.

These herbicides are classified on the basis of

(A) Chemical structure

(B) Mode of action

(C) Time of application

(D) Residual effect

(E) Formulation

(A) On the basis of chemical structure

1. Inorganic : These herbicides do not contain carbon atoms in their molecules (a) Acid type Arsenic acid, Sulphuric acid (b) Salt type Boron, Copper Sulphate, etc.

2. **Organic :** They contain carbon atoms in their molecules. (a) Oil type Diesel oil, polycyclic aromatic oils (b) Non-Oil type Dalaphon, 2, 4-D.

(B) On the basis of mode of action

1. Selective herbicides : The chemical which kills or retard the growth of some plants with little or no injury to other plants. (a) Foliage application

  (i) Contact herbicides directly kill the plant cells e.g., Propanol.

 (ii) Translocated herbicides exert a toxic effect on plant by upsetting plant growth and metabolic processes e.g., 2, 4-D.

  (b) Soil application e.g., chloramber, Trifluralin, etc.

2. Non-Selective herbicides : Herbicides which kill all the vegetation that they come in contact with are called non-selective herbicides.

  (a) Foliage application

      (i) Contact, e.g., Paraquat.

     (ii) Translocated, e.g., glyphosate.

  (b) Soil applications

      (i) Soil Fumigants, e.g., Methyl bromide.

     (ii) Soil sterilants, e.g., Bromacil.

  (c) Aquatic applications used for killing aquatic weeds, e.g., Copper sulphate, Fenac, etc.

(C) On the basis of time of application

  1. Fallow application, e.g., Atrazine.

  2. Pre-sowing application, e.g., Diuron.

3. Pre-emergence application, e.g., Simazine.

4. Post-emergence application, e.g., 2, 4-D.

(D) On the basis of residual effect

1. Short persistent herbicide (up to week), e.g., Paraquat.

2. Medium persistent herbicide (2- 6 weeks), e.g., Butachlor.

3. Long persistent herbicide (Few months), e.g., Alachlor.

(E) On the basis of formulation

1. **Wettable powder:** There forms suspension in water, e.g., Atrazine 80% WP., Simazine 50% WP.

2. **Soluble powders:** Soluble in water, *e.g.*, 2, 4-D, Dalaphon.

3. **Soluble:** Concentrates soluble liquids, *e.g.*, 2, 4-D Amine.

4. **Liquid suspension:** Dissolved in solvents with an emulsifier, *e.g.* Atrazine, Euprazine.

5. **Emulsifiable:** Concentrates an emulsion is one liquid dispersed in another, each maintaining its original identity, e.g., 2, 4-D Ester.

6. **Granules:** There are small pellets and granules, *e.g.*, granules of Butachlor.

RODENTICIDES

Chemical which are used to kill rodents or inhibit or destroy them.

Types

1. Fumigant Aluminium phosphide, Calcium cyanide, Methyl bromide.

2. Anticoagulants Varfarin, Phenyl methyl pyrozolone.

3. Arsenicals Arsenious oxide, Sodium arsenate.

4. Others Zinc phosphide, Calciferol.

NEMATICIDES

Substances that prevent, destroy, repel or inhibit nematodes. There are two types of nematicides

(i) Systemic nematicides (fumigant)

(ii) Contact non fumigant, e.g., Thionazin, Diazinon.

1. Volatile Soil Fumigant Nematicides

These are halogenated hydrocarbons and isothiocyanate groups, e.g., Methyl bromide, Ethylene dibromide (EDB), Methyl isothiocyanate

2. Non-fumigant Nematicides

These are mostly organo-phosphorus and carbamates, available in granular form and can be applied to the rows of the crops at the time of planting or to the soil around the standing trees, e.g., Phorate, Aldicarb Carbofuran, etc.

MOLLUSCICIDES

Substance that prevent repel destroy or inhibit growth of member belonging to phylum Mollusca.

1. Aquatic
    (a) Botanicals sub-group, e.g., Endod
    (b) Chemicals sub-group, e.g., Copper Sulphate, Niclosamide.
2. Terrestrial
    (a) Carbamates Aminocarb, Methiocarb.
    (b) Others Metaldehyde.

Plant Growth Substances

Plant growth regulators

The organic compounds other than nutrients which in small amounts promote inhibit or otherwise modify the plant physiological processes.

They are classified as

(a) Growth promoters, e.g., Auxins, Gibberellins, Cytokinins, Ethylene, etc.
(b) Growth inhibitors, e.g., Abssisic acid, Paraquat.

PHYSIOLOGICAL ROLE

(a) Auxins
- Cell enlargement.
- Root growth.

(b) Gibberellins
Stem elongation.

- Increase in leaf, fruit and blossom size.

- Breaking of dormancy.

(c) Cytokinins

- Cell division.

- Breaking dormancy.

- Growth of stem and roots.

- Chlorophyll synthesis.

(d) Ethylene

- Fruit ripening.

- Flowering.

- Elongation of stem and roots.

(e) Growth retardants

- retard the stem elongation.

- Inhibition of sprouting of tubers.

- Promotion of fruit ripening.

- Increasing the vase life of cut flowers.

- Preventing preharvest fruit drop.

FERTILIZERS

Any commercial chemical compound applied to the soil to supply plant nutrients to crop plant is known as fertilizer.

Types of Fertilizers

Straight Fertilizers : Fertilizers containing and supplying only one of the primary essential nutrients to crop plants are called as straight fertilizers.

TYPES OF STRAIGHT FERTILIZER

1. Nitrogenous fertilizer

(a) Inorganic

(i) Nitrate Fertilizers Nitrogen is combined as NO3 with other elements.

e.g., Sodium Nitrate ($NaNO_3$)   – 16 % N

Calcium Nitrate (Ca (NO_3H)   – 15.5 % N

Potassium Nitrate (KNO_3)   – 13 % N

(ii) **Ammoniacal Fertilizers** Nitrogen is combined in NH4 form with other elements

e.g. Ammonium Sulphate $(NH_4)_2 SO_4$ – 20 % N

Ammonium Phosphate $(NH_4H_2PO_4)$ – 20% N+20% P_2O_5%

Anhydrous Ammonia   – 82% N

(iii) **Nitrate and Ammonium Fertilizers** These fertilizers contain nitrogen in the form of both nitrate and ammonia.

e.g., Ammonium nitrate (NH_4NO_3) – 33% N,

Calcium Ammonium Nitrate   – 26% N

(b) Organic

(i) **Amide Fertilizers** Contain nitrogen in amide form.

e.g., Urea $CO (NH_2)_2$ – 46% N

Calcium cynamide (CaNCN)   – 21% N

(ii) Slow release nitrogenous fertilizer

e.g., Urea- form (urea formaldehyde)   – 38% N

N - Lignin   – 180% N

Nitrate Fertilizers are highly mobile and easily reach to root zone. Due to high solubility in water the leaching losses are more. Ammonium Fertilizers are less rapidly utilized by plants and they are more resistant to leaching losses.

2. Phosphatic Fertilizers

(i) Monocalcium phosphate Ca $(H_2PO_4)_2$

Containing water soluble phosphoric acid

e.g., Single super phosphate   – 16% P_2O_5

Double super phosphate   – 32% P_2O_5

Triple super phosphate   – 48% P_2O_5

(ii) Dicalcium phosphate $Ca_2H_2 (PO_4)_2$

Citric acid soluble phosphoric acid

e.g., Basic slag   – 14-18% P_2O_5

Dicalcium phosphate   – 34-39% P_2O_5

(iii) Tricalcium phosphate Ca (PO$_4$)$_2$

Containing insoluble phosphoric acid

Rock phosphate	– 20-40% P$_2$0$_5$
Raw bone meal	– 20-25% P$_2$0$_5$

3. Potassic fertilizers

(a) Potassium Chloride Fertilizers having potash in chloride form

e.g., Muriate of potash (KCI)   – 60% K$_2$O

(b) Potassium Sulphate Fertilizers having potash in sulphate Form

e.g., Sulphate of potash (K$_2$SO$_4$) – 42-52% K$_2$O

Sulphate of potash magnesia   – 25-39% K$_2$O & 10-12% MgO

(c) Potassium Nitrate Fertilizers having potash in nitrate form

e.g., Potassium nitrate (KNO$_3$)   – 44 % K$_2$O, 13% N

4. Complex fertilizers

Fertilizers containing at least two or more of the primary essential nutrients in chemical combination (N, P$_2$O$_5$ or K$_2$O) are called as complex or multiple nutrient fertilizers.

(i) **Incomplete Complex Fertilizers:** when fertilizers contain only two of the primary nutrients.

(ii) **Complete Complex Fertilizers:** when Fertilizers contain all the three primary nutrients.

(a) Nitro phosphate

Grades (N, P$_2$O$_{5,}$ K$_2$O): (i) 20:20:20 (ii) 15:15:15

(b) **Ammonium Phosphate Sulphate**

Grades (N, P$_2$O$_{5,}$ K$_2$O) (i) 19.5:19.5:0 and 20:20:0

(c) **Diammonium Phosphate**

Grades (N, P$_2$0$_{5,}$ K$_2$O) (i) 18:46:0; (ii) 20:48:0

(d) **Urea ammonium phosphate** Grades (N, P$_2$O$_{5,}$ K$_2$O) (i) 8:28:0 (ii) 20:20:0

5. Fertilizer Mixture

A mixture of two or more straight fertilizer material is referred to as mixed fertilizer (Physical combination).

(a) **Open formula fertilizer mixture:** The fertilizer mixture whose formulae in terms of kind and quantity of ingredients mixed are disclosed by the manufacturers.

(b) **Close formula fertilizer mixture:** The fertilizer mixture whose formulae in terms of kind and quantity of ingredients mixed are not disclosed by the manufactures.

6. Soil amendments

These are the substances that influence the plant growth favourably by changing the plant material in the soil from unavailable to available form, improving the physical conditions of the soil.

Types of Soil amendments

1. Material for correcting soil acidity Lime
2. Material for correcting alkalinity Gypsum

7. Micro-nutrient Fertilizers

The fertilizers supplying the micro-nutrients (essential elements required by crop plants in very small amounts Le. less than 50 ppm) in very minute or trace amounts are called micro-nutrient fertilizers.

Source of Micro-nutrients

1.	Zinc Sulphate	$(ZnSO_4.7H_2O)$	22-35% Zn
2.	Ferrous sulphate	$(FeSO_4.7H_2O)$	20% Fe
3.	Copper Sulphate	$(CuSO_4.5H_2O)$	25-35% Cu
4.	Borax	$(Na_2B_4O_7.10H_2O)$	10.6% B
5.	Ammonium molybdate	$[(NH_4)\ 6M0_7O_{24}.4H_2O]$	54% Mo
6.	Manganese Sulphate	$(MnSO_4.4H_2O)$	23% Mn

This constitutes a very brief overview of the agrochemicals. It may be noted that each of these categories of agrochemicals make a full-fledged topic for a particular discussion, however keeping in view of the requirements of the Seminar, the current overview seems to be sufficient to fulfill the need of the participants.

5

EXPLANATIONS

1. Components of soil

Soils consists of four major components viz., (i) mineral matter, (ii) organic matter, (iii) water and (iv) air. The volume composition of soil in optimum condition for the crop growth is as follows:

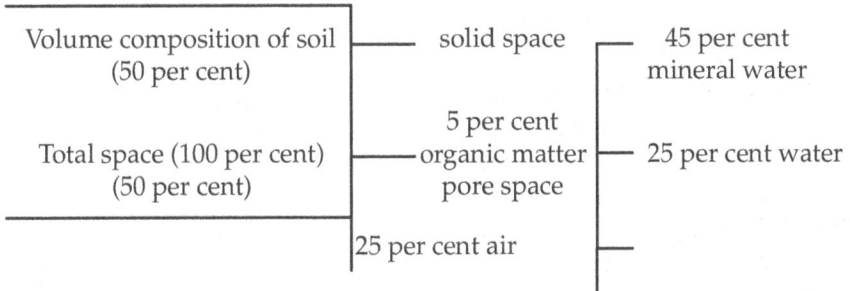

Volume composition of soil (50 per cent)	solid space	45 per cent mineral water
Total space (100 per cent) (50 per cent)	5 per cent organic matter pore space	25 per cent water
	25 per cent air	

Mineral mater in soils: The size and composition of mineral matter in which it has been derived. In general the primary minerals viz., quartz, biotite, muscovite etc. dominate the coarser fractions of soil; on the other hand, the secondary minerals viz., silicate clays and hydrous oxide clays of iron and aluminium etc. are present as very finer fraction, clay in the soils.

Organic matter in soils: The organic matter content in a soil is very small and varies from only about 3 to 5 per cent by weight in a top soil. Organic matter is a store house of nutrients in soil. It is major source of nitrogen, 5-60 per cent of the phosphorus and perhaps about 80 per cent of sulphur. Organic matter acts as a chelate.

Soil water: Soil water plays a very significant role in soil plant growth relationships. Soil water also in soil plant growth relationships. Soil water also presents along with dissolved salts and makes up the soil solution. This soil solution act as an important medium for supplying different essential nutrient elements to growing plants through exchange phenomenon between soil solid surfaces and soil solution and finally between soil solution and the plant roots.

Soil air: Airspace or pore spaces (voids) in a soil consist of that portion of the soil volume not occupied by soil solids, either mineral or organic. Under field conditions, pore spaces are occupied at all times by air and water. Soil air contains various gases like carbon dioxide, very small amount of oxygen and nitrogen etc. soil air differs from the atmospheric air with the relative amounts of those above gases.

2. Surface soil and sub-soil

Surface soil: 1. It is completely weathered. 2. It is dominated by finer particles like silicate clays. 3. Surface soil is porous and friable. 4. Aeration status of surface soil is good and exchange of gases between atmosphere and soil air takes place. 5. The number and activity of soil micro-organisms is very high. 6. Relatively higher organic matter content due to presence of higher biomass on the soil surface. 7. Surface soil has no hard pan. 8. Due to presence of high organic matter content the colour of surface soil is deep brown or dark. 9. It is fertile. Most of the essential plant nutrients are present. 10. It has good physical management condition because of surface soil. 11. Cation exchange capacity (CEC) is very high.

Sub-soil: 1. It is partially weathered. 2. It is dominated by quartz particles and other coarse fragments of minerals. 3. It is more massive and compact. 4. Aeration status of sub-soil is very poor and hence exchange of gases is very much limited. 5. The microbial population and their activity are very low. 6. Due to lack in plant and animal residues in the sub-soil, the amount of organic matter is very low. 7. Sub-soil sometimes has hard pan. 8. The colour of sub-soil is light and sometimes may be light yellowish colour depending on the nature and kinds of unweathered materials. 9. It is less fertile; very few essential plant nutrients are present. 10. It has poor physical condition. 11. Cation Exchange Capacity (CEC) is low.

3. Composition of the Earth's Crust

Rock is composed of elements. Out of 106 elements known, 8 are sufficiently abundant as to constitute about 99 percent by weight of the Earth's crust (upto 16 km). The elements are geochemically distributed into five main groups based on their bonding characters.

(i) **Lithophile elements:** which ionize readily or form stable oxyanions, viz., O, Si, Ti, Fe, Mn, Al, H, Li, Na, K, Rb, Cs, Be, Mg, Ca, Sr, Ba, B, Ga, Ge, Sn, Sc, Y, F, Cl, Br, I, C, HF, Th, P, V, Nb, Ta, Cr, W, U, Zr, (Mo), (Cu), (Zn), (Pb), (Ti), (As), (Sb), (Bi), (S), (Se), (Te), (Ni), (Co) and rare earths.

(ii) **Chalcophile element:** which tend to form covalent bonds with sulphide, viz., S, Se, Ie, (Fe), Ni, Co, Cu, Zn, Pb, Mo, Ag, Sb, (Sn), Cd, In, Ti, As, Bi, Re, (Mn), (Ga) and (Ge).

(iii) **Siderophile elements:** Which readily form metallic bonds, viz., Fe, Ni, Co, Ru, Rh, Pd, Ir, Os and Au.

(iv) **Atmosphile elements:** which tend to remain in atmospheric gases viz., N, (O), He, Ne, Ar, Kr, Xe.

(v) **Biophile elements:** which tend to associated with living organisms, viz. C, H, O, N, P, S, Cl, I, B, Ca, Mg, K, Na, Mn, Fe, Zn,, Cu, Ag, Mo, Co, Ni, Au, Be, Cd, Se, Ti, Sn, Pb, As and V.

4. Occurrence of Soil Forming Rocks

on the basis of their genesis and structure, rocks are generally grouped into three classes, namely (i) Igneous (ii) Sedimentary (iii) Metamorphic. The composition of the upper 5 km of the Earth's crust is as follows:

	Shales	52%	
	Sandstones	15%	
Sedimentary Rocks ⟶			74%
	Limestones and dolomite	7%	
	Granite	15%	
Igneous Rocks ⟶			18%
	Basalt	3%	
Other rocks			8%

5. Relative occurrence of minerals in soils

S.No.	Name of minerals	Percent distribution	Remarks
1	Primary minerals		
	A Ferromagnesians group :		
	(i) Orthosilicates, Ionosilicates pyroxenes, amphiboles and hornblende series. Olivine's	16.8	Weather ability of pyroxenes is medium. Amhiboles are characteristics of metamorphic rocks and medium weather ability.
	(ii) Phyllosilicates or layer silicates. Biotite (black mica) Muscovite (white mica)	3.6	Weather ability of olivihnes is very high. Also biotite very high. In case of muscovite weatherability is low. They are important source of K and clay.
	B Non-ferromagnesium group		
	(i) Tektosilicates	61.0	Weather ability low to very high
	Feldspars	11.6	Very low weathering
	Corthoclase and plagioclases quartz		
	(ii) Secondary minerals clay minerals and Iron oxides etc.		Varied according to the nature of the primary minerals.
	C OthersAccessory minerals	6.0	

6. Important minerals with their chemical composition and weather ability

Mineral	Chemical composition	Weatherability
Quartz	SiO_2	Very low
Zircon	$ZrSiO_4$	Very low
Rutile	TiO_2	Very low
Magnetite	Fe_3O_4	Very low
Ilmenite	$FeTiO_3$	Very low
Orthoclase	$KalSi_3O_8$	Low
Albite	$NaAlSi_3O_8$	Low
Muscovite (white mica)	$Kal_2(AlSi_3O_1)$	Low
Titanite	$CaTiSiO_5$	Low
Garnet	$Ca_3Al_2(SiO_4)_3$	Low
Tourmaline	$NaFe_2Al_4B_2Si_4O_{19}(OH)$	Low
Apatite	$Ca_4(PO_4)_3$, (Ca, F, Cl)	Medium
Serpentine	$Mg_3Si_2O_5 (OH)_4$	Medium
Hornblende	$Ca_3Na (Mg, Fe)_6 (Al, Fe)_{3,} (Si_4O_{11})_4 (OH)_4$	Medium
Augite	$Ca (Mg, Fe)_3 (Al, Fe)_4 (SiO_3)_{10}$	Medium
Calcite	$CaCO_3$	Soluble
Anorthite	$CaAl_2Si_2O_8$	Very high
Olivine	$(Mg, Fe)_2 SiO_4$	Very high
Biotite (black mica)	$K (Fe, Mg)_2 (Al, Fe) (AlSi_3O_{10}) (OH)$	Very high

7. Weather Agents

Physical/mechanical (disintegration)	Chemical (Decomposition)	Biological (Disintegration and Decomposition)
(a) Physical condition of rocks	(a) Hydration	(a) man and animals
(b) Change in temperature	(b) Hydrolysis	(b) Higher plants and their roots
(c) Action of water - fragmentation and transportation	(c) Solution	(c) Micro-organisms
(d) Action of wind	(d) Carbonation	
(e) Atmospheric electric phenomena	(e) Oxidation and reduction	

8. Chemical Weathering

(a) **Solution:** Water is a universal solvent. Most of the minerals water affected by the solvent action of water. Solution is represented by the following typical equation of salt dissolving in water:

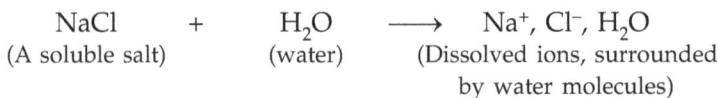

$$NaCl \quad + \quad H_2O \quad \longrightarrow \quad Na^+, Cl^-, H_2O$$

(A soluble salt) (water) (Dissolved ions, surrounded by water molecules)

(b) **Hydrolysis:** Hydrolysis, which involves the splitting of water into H-ions and OH ions, is the reaction of substances with water to form hydroxides and other new substances that are usually more soluble than the original material.

Feldspar + water = clay mineral + soluble cations and anions

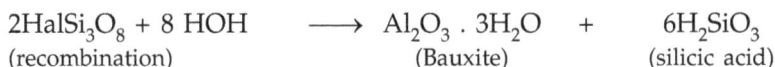

$$KalSi_3O_8 \quad + \quad HOH \quad \longrightarrow \quad HalSi_3O_8 \quad + \quad KOH$$

(orthoclase: very (water) (acid silicate : (very soluble)
slowly soluble) more soluble)

$$2HalSi_3O_8 + 8\ HOH \quad \longrightarrow \quad Al_2O_3 . 3H_2O \quad + \quad 6H_2SiO_3$$

(recombination) (Bauxite) (silicic acid)

Hydrolysis is a double decomposition process.

(c) **Oxidation–Reduction :** Oxidation, as referred to in mineral weathering is both the chemical combining of oxygen with a compound and the loss of electrons (change in oxidation number) of some chemical element. Reduction is the chemical process, in which electrons are gained, the negative charges are increased and the positive charge is decreased. The importance of oxidation-reduction in soil formation is that if can speed mineral breakdown by making some minerals more soluble. Oxidized ions are smaller and higher charged than the ions in the reduced state, so minerals are unstable and weather very rapidly as some of their atoms oxidize. Oxidation and reduction are always companion processes.

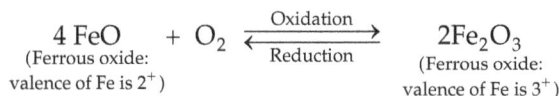

$$4\ FeO \quad + O_2 \quad \underset{\text{Reduction}}{\overset{\text{Oxidation}}{\rightleftharpoons}} \quad 2Fe_2O_3$$

(Ferrous oxide: (Ferrous oxide:
valence of Fe is 2^+) valence of Fe is 3^+)

(d) **Hydration :** Hydration is the chemical combination of a solid substances viz., a mineral or a salt the water. Hydration water combining within the mineral changes the mineral structure, increasing its volume due to swelling and thereby making it softer, more stressed and more easily decomposed.

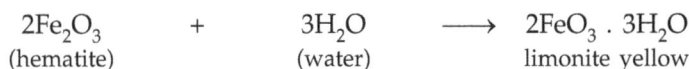

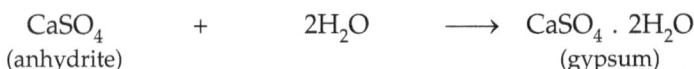

$$2Fe_2O_3 \quad + \quad 3H_2O \quad \longrightarrow \quad 2FeO_3 . 3H_2O$$

(hematite) (water) limonite yellow

$$CaSO_4 \quad + \quad 2H_2O \quad \longrightarrow \quad CaSO_4 . 2H_2O$$

(anhydrite) (gypsum)

(e) Carbonation : Carbonation is the combination of carbon dioxide with any base. This process is very much effective of the other chemical weathering processes. The presence of H^+ ion in percolating waters and other inorganic acids like HNO_3, H_2SO_4 and some organic acids accelerate the decomposition of the minerals through chemical weathering.

The CO_2 gas readily combines with the bases to produce carbonates and bicarbonates as:

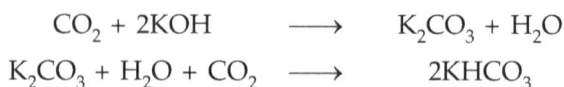

$$CO_2 + 2KOH \longrightarrow K_2CO_3 + H_2O$$
$$K_2CO_3 + H_2O + CO_2 \longrightarrow 2KHCO_3$$

Carbonic acid (H_2CO_3) dissolves minerals more readily than does water alone and forms the soluble bicarbonates as follows:

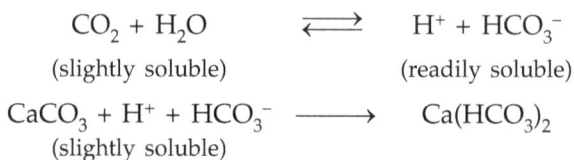

$$CO_2 + H_2O \rightleftharpoons H^+ + HCO_3^-$$
(slightly soluble)   (readily soluble)

$$CaCO_3 + H^+ + HCO_3^- \longrightarrow Ca(HCO_3)_2$$
(slightly soluble)

9. Factors affecting weathering of minerals

Climatic conditions: will tend to control the kind and rate of weathering. Under conditions of low rainfall there is predominant of physical weathering resulting decreasing in their sizes and thereby increases the surface area with little change in volume. On the other hand, the increase in moisture content encourages both chemical and physical changes and thus facilitates the production of soluble and new mineral compounds. Generally the rate of weathering especially chemical weathering is very fast in humid tropical regions.

PHYSICAL CHARACTERISTICS

Composition of Rocks

The phenomenon of breakdown or disintegration encourages due to differential composition of different minerals present in rocks. During the change of temperature the amount of expansion and contraction varies and as a result due to presence of different minerals the differential stress develops and breaks the rocks into mineral component by forming rocks.

Size of the Mineral

The fine sized minerals will decompose more easily and rapidly than that of coarse sized minerals exposing to percolating water. The more chemical reaction will take place in case of finer sized minerals.

Hardness and Cementation

Hardness and Cementation influence weathering primarily by their effect on the rate of disintegration into finer particles, which favours the chemical weathering.

Chemical and Structural Characteristics

The ease of decomposition of mineral depends upon its size, chemical and crystalline characteristics. For example, Gypsum, which is sparingly soluble in water, is dissolved and removed in solution form in the presence of sufficient water. The dark coloured ferromagnesium minerals are more sensitive to chemical weathering than, that of feldspars and quartz owing to their open structures resulting from the positively charged metallic ions of iron and magnesium balancing the negative charge of tetrahedral units.

10. Characteristics of soil separates/particles

Physical nature

(i) **Stone, Gravel and Sands (coarse fragments):** Stones and gavels range in size from 2 mm. Upward and may be almost rounded, irregular angular or even flat. Stone is generally larger (> 3 inches) than gravel (upto 3 inches). Sand (2 – 0.05 mm average particle diameter) may be rounded or quite irregular depending on the mount of abrasion that they have received when sands are not coated with clay or silt particles, such particles do not show any sticky, plastic property or any colloidal property.

(ii) **Clay and silt (Finer fractions):** Clays consists chiefly of secondary products of chemical weathering, have ultramicroscopic size, possessing large surface area than that of silt and sand. Clay particles exhibit properties of swelling, plasticity, cohesion and adhesion etc. Silts are intermediate in size and show properties intermediary between sands and clays and are composted of original mineral fragments.

Mineralogical Nature

(i) **Sands and Silts:** These are coarsest particles among other soil separates. They are fragments of rocks as well as minerals. Quartz (SiO_2) commonly dominates the finer grades of sand as well as silt separates.

(ii) **Clays:** Coarser clay particles are composed of minerals like quartz and the hydrous oxides of iron and aluminium and other aluminosilicate minerals. Three main mineral types' kaolinite, illite and montmorillonite are mostly present.

Mineralogical Nature

Sands and silts: Since quartz (SiO_2) is dominant of these two fractions, they are chemically inactive. Sands contain different insoluble nutrient elements and hence cannot supply nutrients to plants. Silts have been known to release potassium in soils and supply K to the plants.

Clays: This fraction is soils are very active. Montmorillonite and Kaolinite are aluminium silicates. They carry sodium, iron and magnesium. Illiote is hydrous mica, potassium aluminium silicates. It contains high potassium.

11. Determination of textural class

There are generally two methods employed for the determination of textural class.

Feel method: In the field, texture is commonly determined by sense of feel. The soil is rubbed between thumb and finger under wet conditions. Sands feel gritty and its particles can be easily seen. The silt when dry feels like flour and talcum powder and is slightly plastic when wet clavey particles feel very plastic and exhibit stickiness when wet and are hard under dry conditions.

Laboratory method: A more accurate and fundamental method has been devised by the U.S. Department of Agriculture for the naming of soils based on a mechanical analysis. The figure determination of textural class re-emphasize that a soil is a mixture of different sizes of particles.

12. Densities of soil

Density is the weight per unit volume of a substance. It is expressed in gram per cubic centimeter or pound per cubic foot or mega gram per cubic meter.

$$\text{Density (D)} = \frac{\text{Mass (M)}}{\text{Volume (V)}} \text{ gm/cc or lb/cft or mgm}^{-3}$$

Two density measurements – particle density and bulk density are common for soils.

Particle density: The weight per unit volume of the solid portion of soil is called particle density. It is also termed as true density. Generally in the normal soils the particle density is 2.65 g/c.c. or Mg/m^3. The particle density is higher if large amounts of heavy minerals are present with an increase in organic matter of the soil, the particle density decreases.

Bulk density: The mass (weight) per unit volume of a dry soil (volume of solid and pore spaces). It is also expressed in gm/c.c. (C.G.S. System) or lb/cft (F.P.S. System) or mg/m^3. The bulk density of soil is always smaller than its particle density. The bulk density of sand dominated soils is about 1.7 gm/c.c., whereas in organic peat soils the value of bulk density is about 0.5 gm/c.c. Generally in normal soils bulk density ranges from 1 – 1.60 gm/c.c.

13. Porosity of Soil

Pore spaces (also called voids) in soil consist of that portion of the soil volume not occupied by solids, either mineral or organic. The pore space under field conditions, are occupied at all times by air and water. Pore spaces directly control the amount of water and air in the soil and indirectly influence the .ant growth and crop production.

Macropores: large sized pores are referred to as macropores which allow air and water movement readily. Sands and sandy soils have a large number of macropores. It is found in between the granules.

Micro or capillary pores: smaller sized pores are generally referred to as a micro or capillary pores in which movement of air and water is restricted to some extent. Clays and clavey soils have a greater number of micro or capillary pores. It has got more important in the plant growth relationship. It is found within the granules.

Coarse pores: Greater than 0.2 mm or 200 microns (0.008 inch) average diameter.

Medium pores: 0.2 – 0.02 mm, which is 200–20 microns (0.008–0.0008 inches) average diameter. Size of coarse silt particles.

Fine pores: 0.02 – 0.002 mm, which is 20–2 microns, (0.0008 inch) average diameter. Size of fine silt particles.

Very fine pores: Less than 2 microns (0.00008 inch) average diameter. Size of large clay particles.

$$\% \text{ pore space} = 100 \left(1 - \frac{\text{bulk density}}{\text{particle density}} \right)$$

14. Reasons for poor aeration

There are generally two reasons by which poor aeration results.

Excess moisture: When soil is subjected to excess moisture waterlogged condition is developed. This situation is generally found on poorly drained, fine textured soils having a minimum of macropores through which water can move very rapidly. In such condition most of the plants cannot grow. Such poor aeration can be prevented through the removal of excess water either by drainage or by controlled run off.

Gaseous interchange: The adequate interchanges of gases between the soil and free atmosphere depend on two factors:

The rate of biochemical reactions influencing the soil gases, and

The actual rate at which each gas is moving into or out of the soil. The exchange of gases between the soil and the atmosphere is facilitated by two mechanisms.

(a) Mass flow : Mass flow of air is apparently due to pressure differences between the atmosphere and the soil air. The temperature may influence the renewal of soil air by two ways: There may be temperature variations within the soil between the different horizons. The contraction and expansion of the air within the pore spaces as well as the tendency for warm air to move upward may cause some exchange between the different layers and with atmosphere.

The soil and the atmosphere usually have different temperatures. This differential temperature also permits an exchange between the atmosphere and soil air in the immediate surface.

(b) Diffusion: Diffusion is the molecular transfer of gases. The molecules of gases are in a state of movement in all directions. Through this process each gas tends to move in a direction determined by its own partial pressure. Diffusion allows movement from one area to another. Diffusion process seems to be directly related to the volume of pore spaces filled with air.

15. Composition of soil and atmospheric air

Name of gas	Percentage by volume	
	Soil air	Atmospheric air
Oxygen	20.00	21.00
Nitrogen	78.60	78.03
Carbon dioxide	0.50	0.03
Argon	0.90	0.94

16. Oxygen diffusion rate (ODR)

Techniques for measurement of ODR

Principles: When a certain electric potential applied between a platinum electrodes inserted into the soil and a reference electrode, oxygen (O_2) is reduced at the platinum electrode surface. The general reaction taking place at the platinum micro-electrode surface in the reduction of O_2 is in two steps involving two electrons in each step. The reactions in two different media are

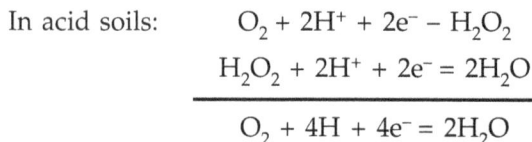

In acid soils:

$$O_2 + 2H^+ + 2e^- - H_2O_2$$

$$H_2O_2 + 2H^+ + 2e^- = 2H_2O$$

$$\overline{O_2 + 4H + 4e^- = 2H_2O}$$

In neutral or alkaline soils :

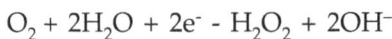

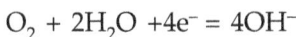

$$O_2 + 2H_2O + 2e^- - H_2O_2 + 2OH^-$$
$$H_2O_2 + 2H^+ + 2e^- = 2H_2O$$

$$O_2 + 2H_2O + 4e^- = 4OH^-$$

An electric current flows between the two electrodes and is proportional to the rate of O_2 reduction the rate of O_2 reduction is in turn related to the rate at which it diffuses to the electrode.

$$ODR = \frac{it \times 10^{-6} \times 60 \times 32 \times 10^{-6}}{4 \times 96,500 \times A} \mu gcm^{-2} min^{-1}$$

where,

n = 4 $\longrightarrow$ no. of electrons required to reduce the one molecule of O_2

F = Faraday constant (96,500 coulombs)

A = Area of the electrode surface (cm^2)

17. Sources of soil heat

Solar radiation: Radiant energy from the sun is the power source that determines the thermal regime of the soil and he growth of plant. The angle at which the sun's rays meet the earth greatly influences the amount of radiation received per unit area. Radiation reaching the earth at an angle is scatted over a wide area than the same radiation striking the earth's surface perpendicularly. So, solar radi.

Bio-chemical reactions: In the soil atmosphere a variety of chemical reactions are going on and during such reaction liberation of large amount of heat in the soil environment results. Besides decomposition of organic matter and other crop residues in the soil and other microbial processes liberate large amounts of heat in the soil and thus contributes soil heat.

Conduction: The inner atmosphere of the earth is very hot; the conduction of the heat to the soil environment is very slow. Generally, during night, the uppermost surface soil becomes cooler than sub-surface soil. Thus, heat flows from the regions of subsoil to the region of surface soil cooler soil layer.

Precipitation: During the winter season precipitation increases soil heat because of its (precipitation) higher specific heat.

Exposure: Exposure is little importance in tropics because of the high elevation of the sun. It is of significance in the middle latitude where the elevation is lower.

Vegetation: Vegetation plays a significant role of soil heat because of the insulating properties of plan cover. Bare soil is unprotected from the direct rays of the sun and becomes very warm during the hottest part of the day.

18. Influence of soil temperature on plant growth and nutrition

Germination of seeds: If the temperature is too low, the seed fails to germinate at a slow rate. Seeds may be injured if the temperature will be very high.

Physical properties of soil: The temperature has a great influence on soil structure, aggregation of the soil as well as on the binding materials present in it.

Microbial activity : The activity of micro-organism shaving thermophobic and thermophilic nature is influenced by the variation in soil temperature is below 5°C and above 54°C. The optimum temperature for the activity of most of the micro-organisms is in the range of 25-35°C.

Decomposition of organic matter in soil : At low temperature the rate of organic matter decomposition is low resulting various toxic organic substances in soil and the high temperature the rate of the same is very fast resulting beneficial products of organic matter decomposition and hence influence the plant growth.

Variation in soil temperature (very low to high temperature) affects the absorption of soil water by the plant roots.

Availability of nutrients: Various physico-chemical and chemical reactions are greatly influenced by soil temperature. Temperature influences the solubility reactions of different nutrients and releases larger amount of nutrient elements in the soil solution at high temperature.

Root growth: Low temperature encourages white succulent roots with little branching, while high temperature encourage a browner, finer and much more freely branching root system.

Plan diseases: Development of various diseases is also related to the soil temperature. At low temperatures, the soil contains many weekly parasitic fungi which will grow actively and very rapidly and so those will kill the seedlings. Seedlings of the temperature zone cereals, which are adapted to grow actively at the lower temperature, are relatively resistant to their attack.

19. Determination of soil colour

The soil colours are best determined by the comparison with the Mussel colour chart. The colour of soil is a result of the light reflected from the soil. Soil colour rotation is divided into three parts:

Hue: It denotes the dominant spectral colour (red, yellow, blue and green).

Value: It denotes the lightness or darkness of a colour (the amount of reflected light).

Chroma: It represents the purity of colour (strength of the colour).

The Munsell colour notations are systematic numerical and letter designations of each of these three variables (Hue, Value and chroma).

20. Measurement of soil moisture

Soil moisture can be measured by following methods: Gravimetric method: This method is the classical procedure used as the check for all other methods. A soil is sampled, put into a container, weighted in a sampled (moist) condition, oven dried, and weighed against after drying. Drying is done at 105°–110° (221°–230°F) to constant weight.

$$\% \text{ moisture} = \frac{\text{Weight of moist or wet soil} - \text{weight of oven dry soil}}{\text{Weight of oven dry soil}} \times 100$$

Electrical conductivity method: This method is based upon the changes in electrical conductivity with the variation in the soil moisture. The gypsum block inside of which are two electrodes at a definite distance apart are used. The blocks are buried in the soil at desired depth and the conductivity is measured with a modified Wheatstone bridge from this method the percentage of moisture from the field capacity to the wilting percentage can be easily measured.

This method cannot be used in soils contained high salt concentrations, which interferes during the measurement. This method is practiced in the laboratory.

Measurement by using Tensiometers: Tensiometers measure the metric potential of soil moisture in situ (filed) with the use of a porous clay cup attached to a tube filled with water. The water in the cup and tube is attached to a vacuum gauge or a mercury manometer. As the soil dries, water moves out through the porous cup, creating a suction or vacuum on the water column. These suction readings are then calibrated on the gauge to a readings are then calibrated on the gauge to a specific soil to interpret the percent of moisture tensiometers are used to schedule irrigation. Tensiometers do not measure soil metric potential values as low as the usual wilting values. The actual range of effective measurement is only from 0 to – 0.85 bars.

Neutron scattering method: Neutron probes look like flashlight cylinders with long cords attached. The probe contains radioactive material (radium or beryllium) that emits rapidly moving neutrons. As the neutrons emitted from the probe collide with hydrogen ions (of which water is a major source), they are slowed and deflected, and some of the slowed deflected neutrons are deflected back to the probe where a counter measures them. Only slowed neutrons are counted.

21. Atterberg Limits

Upper plastic limit: It is also called liquid limit. It represents the moisture content of soil at a point where the soil water mass just flows under an applied forced and fails to retain its shape.

The upper plastic was determined originally by placing a small amount of soil in a round bottomed dish, working it into a stiff paste, pressing it tightly against the bottom, cutting a 'V' shaped groove in the plastic mass, and jarring the dish to make the two segments flow together. The cut and dry process was repeated until the correct flow was obtained. The moisture content of the plastic soil was then determined.

Lower plastic limit: refers to the moisture content of a soil at a point where its consistence changes from plastic to triable and the soil-water mass is unable to change shape continuously under the influence of an applied force. And ultimately the mass breaks into fragments.

Plasticity Number (Index): It refers to the difference between moisture contents of soil at its upper plastic limit and lower plastic limit. Different soils are characterized by a specific plastic number or index of plasticity.

22. Nature and some important properties of soil colloidal particles

Adsorption: The surface adsorption is very large when large amount of colloidal materials are present in a substance because of having large surface area contributed by its presence. The adsorption of ions is governed by the type and nature of ion and the type of colloidal particle.

Brownian movement: Colloidal particles are found to be in continual motion. The oscillation is due to the solution of the colloidal particles or molecules of the liquid in the dispersion medium. This movement is primarily responsible for the coagulation or flocculation of soil colloidal materials. (clay and humus).

Electrical charge: Colloidal particles usually have an electrical charge-some positive and some negative when clay colloids suspended in water, it carried a negative electric charge and thereby attracts positively charged ions (cations).

Flocculation: The colloidal particles are coagulated by adding an oppositely charged ion. This process of formation of flocks in known as flocculation. As for of formation of flocks is known as flocculation as for example, clay is coagulated by the use of aluminium (Al^{3+}). This condition is generally beneficial in relation to Agriculture since it is the first step in the formation of sable aggregates or granules. The ability of common cations to flocculate soil colloids is in the order of,

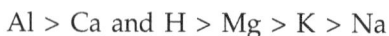

Al > Ca and H > Mg > K > Na

Plasticity: Soils containing more than 15 per cent of colloidal clay exhibit plasticity. Plasticity occurs only when soils are moist or wet. Plasticity phenomenon is extremely important because it encourages a change in soil structure which is most related to the tillage operations.

Cohesion and adhesion: Due to cohesion force clay particles are able to form aggregates and also due to adhesive force clay particles envelope sand particles. These to forces help in the retention of water in the soil as well as absorption to the plants and micro-organisms.

Swelling and shrinkage: contraction and expansion in volume of soil with the variation in moisture are influenced by the presence of colloidal clay particles in soil. Swelling phenomena occurs when colloidal clay particles are allowed to be placed in contact with moisture due to imbibitions of water. Whereas shrinkage occurs in dry condition of the soil containing sufficient amount of colloidal particles and causes decrease in volume with cracks in variable nature.

Non-permeability: Colloids are unable to pass through a semi-permeable membrane and this membrane allows passing water and other dissolved substances but retains colloidal materials.

23. Sources of negative charges on silicate clays

There are generally two types of charges, i.e., one p^H independent originate from exposed crystal surfaces and isomorphous substitution respectively.

Exposed crystal edges: The negative charge on silicate clays develops due to unsatisfied valences at the broken edges of silica and aluminium sheet. Besides the flat external surfaces of silicate clay minerals also serve as the source of negative charge. Reactions are as follows:

$$SiOH + OH^- \rightleftharpoons SiO^- + H_2O$$

$$AlOH + OH^- \rightleftharpoons AlO^- + H_2O$$

Isomorphous substitution: Isomorphous substitution is the substitution of one ion for another of similar size but lower positive valence. However, when a magnesium (Mg^{2+}) ion about same diameter as an aluminium ion (Al^3) replaces one of the aluminium by isomorphous substitution, an imbalance occurs. Consequently, the aluminium octahedral sheet assumes one negative charge for each Mg^{2+} for Al^{3+} substitution. The charges arise from such isomorphous substitution are not dependent on p^H and therefore, these charges are commonly referred to as permanent or p^h independent charges.

Anion exchange: Some clay minerals exhibit positive as well as negative charges. This will cause anion exchange between hydroxyl ions (OH) and anions like phosphate, sulphate, chloride etc.

24. Mohr and van Barren recognized the following five stages of soil development

Initial stage: Unweathered parent material.

Juvenile stage: Weathering just started, but much of the original material is still unweathered.

Virile: Easily weatherable minerals have been decomposed for the greater part, the clay content has increased and certain mellowness is discernible. The content of soil components less susceptible to weathering is still appreciable.

Senile: Decomposition arrives at a final stage, and only the most resistant minerals to weathering have survived.

Final: Soil developed has been completed and the parent material is completely weathered.

25. Soil orders in the 7th Approximation

Entisols: little horizonation because of parent material that is very resistant to weathering. Soils of this order are found under a wide variety of climatic conditions.

Sub-orders: Aquents, Arents, Psammens, Fluvents.

Vertisols: Little horizonation because of extreme argillipedo turbation. The central concept of vertisols is that of clavey soils that has deep wide cracks at some time of the year and has high bulk density between the cracks. Soils of this order are sticky and plastic.

Sub-orders: Xererts, Torrets, Uderts, Usterts.

Inceptisols: Soils of this order have mainly diagnostic horizons that form rapidly. Rare in desert because of moisture regime restrictions soils of this order is generally found in south-western India and along the Ganga's rivers.

Sub-orders: Aquepts, Andepts, Plaggepts, Tropepts, Ochrepts.

Aridisols: These mineral soils are mostly found in dry climates. They have an ochric epipedon generally light in colour and low inorganic matter. They have a horizon of calcium carbonate (calcic), gypsum (gypsic) or even more soluble salts (salic). Aridisols may be productive if irrigation water is available sub-Orders: Orthids, Argids.

Mollisols: Mollisols are extensive and important agricultural soils. These are very dark coloured, base rich soils. They have developed in lime rich parent material in which there has been decomposition and accumulation of large amounts of organic matter. The formation of mollisols is favoured by semi-arid of sub-humid climates, but mollisols, particularly Udolls (Sub-order of Mollisols) are in more humid regions. The native vegetation is usually grasses Mollisols have the mollic epipdon (GK. epi = over; and pedon = soil, diagnostic surface horizon).

Spodosols: These soils have been designed in soil Taxonomy to include those soils that were classified as podzols and Groundwater Podzols of earlier classification systems. These are soils with accumulation of amorphous materials (humus and sesquioxides) in subsurface horizons. They are characterized by a bleached, wood ash-coloured A_2 horizon and an illuvial horizon of humus and free sequioxide accumulation. Most spododols are coarse textured low in phyllosilicate clay, and naturally quite acid in addition to having a spodic or placic horizon.

Alfisols: Alfisols are soils with grey to brown surface horizon, medium to high base supply, and subsurface horizons of clay accumulation (Argillic), usually moist but may be dry during warm season. An alfisoll may also have tragipan, duripan, nitric horizon, petrocalcic horizon, plinthite or other featuers and these feathers are used in defining the various great groups within the order. The soils have marks of process that translocate silicate clays without excessive depletion of bases and without dominance of the processes that lead to the formation of a moollic epipedon. Some of he red and lateritic soils of India have been classified as Alfisols.

Ultisols: The ultisols are soils of mid to low latitudes that have argillic or kandic horizon with a low base supply. They are characterized by low (<35%) base saturation, and accumulation of clay in the sub-soil. Base saturation in most ultisols decreases with depth because the vegetation has cycled the bases. The ultisols are most extensive in warm humid climates that have a seasonal deficit of precipitation. Because the low fertility and low base status, these soils pose limitations for agricultural use. Generally these soils are under forests.

Oxisols: Oxisols are soils with oxic horizons or a few candic horizons that meet the weatherable minerals requirements of the oxic horizons. They are the most highly weathered soils. Oxic horizons have a very low cation exchange capacity per unit of clay and are very low in fertility. In India these soils are generally found in Kerala, Tamil Nadu, Karnataka and Orissa states.

Histosols: Histosols are organic soils composed mainly of high quantities of plant, but also sometimes of animal residues in various stages of decomposition. These soils are commonly called bogs, moors or peats and mucks. Most Histosols are saturated or nearly saturated with water most of the year unless they have been drained. Histosols, contain at least 12 or 18 percent organic carbon saturated with water and 20 percent or more organic carbon when not saturated with water.

Andisols: Soils of this order developed from volcanic eruption. Incomplete weathering of geological materials in found. The soils of surface layer are dark in colour with low bulk density. During weathering of volcanic materials, amorphous or poorly crystalised minerals like allophane, imogolite and ferrihydrite are produced on the surface layer.

Gelisols: It is a recently introduced soil order which accommodate soils exhibiting the evidence of cryoturbation (frost mixing or churning) through which this soil has developed. It is still not recognized in India, but may be found in higher altitude of Himalayas like Jammu and Kashmir and Sikkim with perma frost conditions.

26. Diagnostic Horizons and their characteristics

(A) **Epipedons:** (surface horizons)

(i) **Mollic (Mollis = soft):** Strong soil structure, thick dark coloured, 50% or more base saturation, organic carbon not less than 0.6%, less than 250 mm 1% citric acid soluble phosphate, does not have the high water contents of sediments that have been continuously under water since deposition beneath water n value is (0.7).

(ii) **Umbric (Umbra shade/dark):** Same as mollic epipedon except for base saturation (<50%).

(iii) **Anthropic (Anthrops = man):** Formed due to long term effect of human use of the soil for residence purposes, more citric acid soluble P_2O_5 and all other characteristics are same as mollic epipedons except moisture status.

(iv) **Histic (Hostos = tissue):** Very high inorganic carbon content, usually found in very poorly drained mineral soils that are close to organic soils.

(v) **Plaggen (Plaggen = sod):** Formed due to long continued manuring where sod was used as bedding for animals in stables and borns and subsequently was deposited on other soils. Thickness is 50 cm or more.

(vi) **Ochric (Ochros: Pale, light coloured):** Light coloured low organic carbon content, may be hard and massive when dry.

(B) **Endopedons (sub-surface horizons):** Agric (Ager = field): organic and clay accumulation just below the plough layer, C/N ratio in the agric horizon is low (usually < 8), p^H is close to neutrality (6 – 6.5), illuvial materials (humus and clay) accumulate as lamelac directly below the very surface horizon.

Albic (Albus = white): It is defined as a surface or lower horizon that has such thin or discontinuous coatings on the sand or silt particles that the hue and chroma of the horizon are determined chiefly by the colour of the sand and silt particles.

Argillic (Argilla = white clay): Silicate clay accumulation, an argillic horizon should be at least one tenth as thick as the sum of the thickness of all overlying horizons formation of argillic horizon is slow, it is not genetically young.

Calcic (Calcic = lime): It has two forms (i) underlying materials have less carbonate than the calcic horizon and it includes horizons of secondary carbonate environment that are 15 cm thick or more and (ii) calcic horizon is 15 cm or more thick, has a $CaCO_3$ equivalent ≥5 per cent by volume of identifiable secondary carbonates as\s pendants on pebbles, concretions.

Natric (Natrium = sodium): It is a special kind of argillic horizon, high in sodium (SAR e•13), more exchangeable magnesium plus sodium than calcium plus exchange acidity (at p^H 8.2), columnar or prismatic structure.

Spodic (spodos = wood ash): The spodic horizon is one in which a significant quantity of active (high CEC, large surface area, high water retention) materials composed of organic matter and aluminium with or without iron have precipitated and dominate reactive properties of the soil material.

Oxic (for oxides): The oxic horizon is at least 30 cm thick consisting of a mixture of hydrates oxides of iron or aluminium, or both, with variable amount of kaolinite and necessary highly insoluble minerals such as quartz, sand.

(C) Hardpan Horizons: Duripan (durus = hard, plus pan = hard pan): A sub-surface horizon that is cemented mostly by silica. Although carbonates may be present, duripans would not slake in water nor in 18 per cent hydrochloric acid (HCl) but will distintegrade in hot concentrated potassium hydroxide (KOH) solution or alternating acidic and basic solutions, both of which dissolve silica.

Fragipan (modified from fragilus = brittle, and pan= brittle pan): A natural sub-surface horizon with high bulk density relative to A and B horizons (the solum) above, seemingly cemented when dry but showing a moderate to weak bittelness when moist. Low in organic matter. Base saturation and p^H of soil are low.

Plinthite (plinthos = brick): It is iron-rich, humus poor moisture of day with quartz and other diluents. It commonly occurs as dark red mottles, which usually are in platy, polygonal or reticulate patterns.

27. Types of soil survey

Detailed soil survey: In this survey boundaries of soil units are delinated from observations by actual traverses throughout the course of the boundary. Detailed soil surveys are conducted to furnish information required for a proper assessment of soil properties, terrain features, errosional aspects and other related factors that can help in working out the use capability and management practices for soil conservation and better production of crops and maintenance of soil fertility. Cadastral maps (1:8000 or 1: 4000 scale) or aerial photographs (1:15,000 scale) are generally used as base material for preparing soil maps for detailed soil survey. These surveys are laborious time consuming and much expensive. Detailed soil survey is of two types, i.e., low and high intensity survey.

Reconnaissance soil survey: This type of survey is undertaken to prepare resource inventory of large areas. It identifies broadly the kinds of soils and their extent of distribution. In these surveys the soil boundaries are not totally traversed, but drawn partly by extrapolation. The scale of mapping is 1:50,000 using topographical maps of the survey of India as base material or aerial photographs of similar scale wherever available. Reconnaissance soil surveys give information for detailed soil surveys and broad land use planning and agricultural development.

Detailed–Reconnaissance soil survey: It is a combination of reconnaissance and detailed soil surveys and is undertaken for understanding distribution of basic soil classes of series and their phases.

Semi-detailed soils survey: This kind of soil survey comprises very detailed study of some selected strips cutting across much overial photo interpretation (API) units for developing correlation between API units and soils. This type of soil survey provides adequate information about various, soil including problematic soils.

Recently there are two other types of soil surveys have been recognized i.e., exploratory and rapid reconnaissance soil survey. These lead to preparation of small scale soil maps that are needed for macro level planning for diversified agro-based development programmes.

28. Land Capability Classes

It consists of eight classes viz., Class I to Class VIII, Classes I through IV can be used for cultivation and classes V through VIII cannot be cultivated in their present state under normal management.

Class I	very good land	
Class II	good land	land suitable for cultivation
Class III	moderately good land	
Class IV	fairly good land	
Class V, VI and VII	Land suitable for pastures and grazing	land suitable for cultivation
Class VIII	Land suitable for wild life and watershed	

29. Importance of buffering in agriculture

The importance of buffering of soils is mainly two folds: the stabilization of soil p^H and the amounts of amendments necessary to affect a certain change in soil reaction or soil p^H.

An abrupt charge in p^H causes a radial modification in soil environment and if it fluctuates too widely, the plants and micro-organisms would suffer seriously. Changes in soil reaction not only have a direct influence on the plants but also exert an indirect influence on soil environment by creating certain sudden charges in the availability of nutrients. Deficiencies of certain plant nutrients and excess of others in toxic amounts would seriously upset the nutritional balance in the soil.

The amounts of amendments necessary to affect a given alteration in soil reaction also relate the effectiveness of p^H stabilization. The greater the buffering capacity of the soil the larger must be the amounts of lime or sulphur to affect a given change in p^H.

30. Sources of soil acidity

Leaching due to heavy rainfall: Generally acid soils are common in all regions where rainfall or precipitation is high enough to leach appreciable amounts of exchangeable bases from the surface soils and relatively insoluble compounds of Al and Fe remains in soil. The nature of these compounds is acidic and its oxides and hydroxides react with water (H_2O).

Acidic parent material: Some soils have developed from parent materials which are acid, such as granite and that may contribute to some extent soil acidity.

Acid forming fertilizers and soluble salts: The use of ammonium sulphate $(NH_4^+)_2SO_4$ when applied to the soil replace calcium (Ca^{2+}) ions from the exchange complex and the calcium sulphate ($CaSO_4$) is formed and finally leached out. Divalent cations of soluble salts usually have a greater effect on lowering soil p^H than monovalent metal cations.

Humus and other organic acids: Humus materials in soils occur as a result of microbiological decomposition of organic matter and contain different functional groups like carboxylic (–COOH), phenolic (–OH) etc. which are capable of attracting and dissociating hydrogen (H^+) ions. During organic matter decomposition humus organic acids and different acid salts may also be produced and increased total acidity of soils.

Aluminosilicate minerals: At low p^H values most of the aluminium (Al) is present as the hydrated alminium ions (Al^{3+}) which undergoes hydrolysis and release hydrogen (H^+) ions in the soil solution.

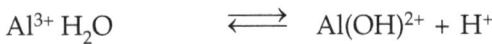

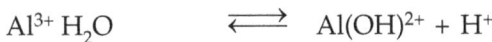

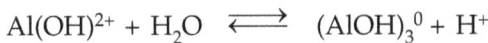

$$Al^{3+}\ H_2O \rightleftharpoons Al(OH)^{2+} + H^+$$

$$Al^{3+}\ H_2O \rightleftharpoons Al(OH)^{2+} + H^+$$

$$Al(OH)^{2+} + H_2O \rightleftharpoons (AlOH)_3^0 + H^+$$

$$(AlOH)_3^0 + H_2O \rightleftharpoons (AlOH)_4^- + H^+$$

It is also possible that structural OH⁻ ions at corners and edges may dissociate hydrogen (H^+) ions and develop soil acidity.

Carbon dioxide (CO_2): Soil containing high concentration of CO_2 which ultimately helps the soil to become acidic.

Hydrous oxides: These are mainly oxides of iron and aluminium. Under favourable conditions they undergo stepwise hydrolysis with the release of hydrogen (H^+) ions in the soil solution and develop soil acidity.

Aluminium and Iron polymers: The Al^{3+} ions displaced from clay minerals by cations are hydrolyzed to monomeric and polymeric hydroxy aluminium or iron forms are illustrated by stepwise reactions with the liberation of hydronium ion (H_3O^+) and lowe soil p^H.

Kinds of soil acidity

Active Acidity: Active acidity may be defined as the acidity develops due to hydrogen (H^+) and aluminium (Al^{3+}) ions concentrations of the soil solution. The magnitude of this acidity is limited.

Potential reserve/exchange acidity: Exchange acidity may be defined as the acidity develops due to absorbed hydrogen (H^+) and aluminium (Al^{3+}) ions on the soil colloids. The magnitude of this acidity is very high. An equilibrium relationship between exchange and active acidity on 2:1 type colloid.

Total acidity: active acidity + Exchange acidity

31. Problems of soil acidity

It may be divided into 3 groups.

(i) Toxic acidity : Acid toxicity: The higher hydrogen ion concentration is toxic to plants under strong acid conditions of soil the acid toxicity includes possible toxicities of acid anions as well as H^+ ions.

Toxicity of different nutrient elements: Iron and manganese: The concentration of Fe^{2+} and Mn^{2+} in soil solution depends upon the soil reaction or p^H, organic matter and intensity of soil reduction. As a result of soil reduction, the Mn^{4+} and Fe^{3+} reduce to Mn^{2+} and Fe^{2+} respectively and increase their concentrations to a very high and toxicity of those elements develops. Due to such toxic effects, a physiological disease of rice is found in submerged soils which are popularly known as browning disease.

Aluminium: It restricts the root growth.

It affects various plant physiological processes like division of cells formation of DNA and respiration etc. It restricts the absorption and translocation of some important nutrient elements form soil.

It causes wilting of plants.

It also inhibits the microbial activity in the soil.

(2) Nutrient availability: Non-specific effects: It is associated with the inhibition effect of root growth and thereby affects the nutrient availability.

Specific effects: Deficiency of exchangeable bases like Ca^{2+} and Mg^{2+} are found in acid soils.

In acid soils Fe, Mn, Cu and Zn are abundant but Mo is very limited and unavailable to plants. In acid soils having very low p^H the availability of boron may also be decreased due to absorption on sesquioxides, iron and aluminium hydroxy compounds. Nitrogen, potassium and sulphur become less available.

Microbial activity: It is well know that soil organisms are influenced by fluctuation in the soil reaction. Fungi can grow well under very acidic soils and caused various diseases like root rot of tobacco, blights of potato etc.

32. Effect of over liming

When excessively large amounts of lime applied to an acidic soil the growth of plant is affected. Deficiency of Fe, Cu and Zn will occur.

Phosphorus and potassium availability will be reduced.

Due to high OH^- ion concentrations by over liming, root development will be inhibited is association with tip swelling brought about by hydrations.

Due to over liming, boron deficiency will occur.

The incidence of diseases like scab in root crops will be increased.

33. Influence of lime on soil properties in relation to plant nutrition

(a) Direct benefits: Toxicity of Al and Mn can be reduced by the application of liming materials. The reduced uptake of Ca^{2+} and Mg^{2+} in the soil solution can also be alleviated with the application of lime. Removal of hydrogen (H^+) ion toxicity which damages root membranes and also causes detrimental effect for the growth of beneficial micro-organisms.

(b) Indirect effects: The application of liming materials to acid soils will inactivate the iron and aluminium thus increasing the level of plant available phosphorus.

Fe, Mn, Zn, Cu, B and MO availability increases with a decrease in soil pH the toxic effect of most of the micronutrients can be prevented by the application of lime.

The process of nitrogen fixation both symbiotic and non-symbiotic is favoured by adequate liming.

The structure of fine textured soil may be improved by liming.

Club root disease of Cole crops can be reduced with the application of lime.

Liming increases the efficiency of different fertilizers especially nitrogenous and phosphatic fertilizers by modifying the soil reaction favourably.

34. Sources of soluble salts

Primary minerals: It is the original and important source of all the salt constituents. During the process of chemical weathering, which involves hydrolys is hydration, solution, oxidation and carbonation, various constituents like Ca^{2+}, Mg^{2+} and Na^{2+} are gradually released and made soluble.

Arid and semi-arid climate: Salt soils are mostly formed in arid and semiarid regions. Where low rainfall and high evaporation prevails. In dry weather, the salts move up with the water and brought up to the surface where they are deposited as the water evaporates.

Ground water: It contains large amount of water soluble salts which depends upon the nature and properties of the Geological material with which water remains in contact where water table and evapotranspiration rate is high, capillary activity and the salts accumulate on soil surface in the form of crystallization.

Ocean or sea water: Sea water enters into the land by inundation and deposited on the soil surface as salts.

Irrigation water: The application of irrigation water without proper management (i.e. lack of drainage and leaching facilities) increases the water table and surface salt content in the soil.

Use of basic fertilizers like sodium nitrate ($NaNO_3$) basic slag etc. may develop soil alkalinity.

35. Problems of salt affected soils

(a) Saline soils: Soils are usually barren but potentially productive.

Wilting coefficient is very high amount of available soil moisture is low. During a dry period, slat in soil solutions may be so concentrates as to kill plants by pulling water from them (exoosmosis).

When salt concentration increases it produces toxic effects directly to plants such as root injury, inhibition of seed germination.

(b) Alkali or sodic soils: Dispersion of soil colloids: The influence of exchangeable sodium on the overall physical properties of soils is associated mainly with the behaviours of clay and organic matter, in which most of the cation exchange capacity (CEC) is concentrated. The sodium ions adsorbed by clay colloids causes deflocculation or dispersion of clay which result in a loss of desirable soil structure and helps for the development of compact soil.

Due to dispersion and compactness of soil, aeration, hydraulic conductivity, drainage and microbial activity are reduced.

Caustic influence: It results high sodicity caused by the sodium carbonate (Na_2CO_3) and bicarbonate ($NAHCO_3$).

High hydroxyl (OH^-) ion concentration no doubt has direct detrimental effect on plant Damage from hydroxyl ions occurs at pH 10.5 or higher.

The presence of excess sodium in sodic soil may induce deficiencies of other cation like calcium (Ca^{2+}) and magnesium (Mg^{2+}).

The high pH in alkali or sodic soils decreases the availability of many plant nutrients like P, Ca, N, Mg, Fe, Cu, and Zn.

36. Criteria for evaluation of irrigation water

Salinity hazard or total concentration of soluble.

Sodium hazard or relative sodium concentration

Salt index

Bicarbonate hazard residual sodium carbonate (RSL).

Boron concentration

Chloride concentration

Soluble sodium percentage (SSP)

Magnesium hazard

Nitrate concentration

Lithium

37. Optimum conditions for soil microbial activity

Moisture: Dryness kills many microbes and many micro-organisms develop resistant strains or enter a dormant stage. Micro-organisms are less active at the wilting percentage for plants or at the other extreme saturated soil.

Soil reaction: At neutral soil reaction (pH 7.0) soil micro-organisms growth is high. Localized micro-nutrients near roots or decomposing residues can produce localized lower pH than that of the whole soil.

Temperature: The activity of most of soil organisms accelerates with the rise in soil temperature. There are 3 groups based on temperature.

Psychrophiles: Grow at temperature <5°C.

Mesophiles: Grow slightly near 0°C (32°F). Optimum temperatures are usually between 25°C and 37°C (77-90°F).

Thermophiles: This group of organisms can tolerate 45° – 75°C (113°-167°F) with optimum between 5° and 65°C (131° –149°F).

Micro-organisms have greater demands for nutrients for their growth and development, especially for N, P, S and Ca, so the optimum amount of those nutrients should be present in the soil.

Soil having moderate to high organic matter is most favourable for the optimum growth and development of most of the soil organisms.

Some soil antibodies substances the retards or kill other beneficial micro-organisms in the soil.

Many organisms act as predators and parasites. Sometimes they kill many of beneficial soil micro-organisms in soil and indirectly inhibit many important transformations.

38. Function of soil micro-organisms

Beneficial: Organic matter decomposition. By this process, plant and animal residues are broken down by micro-organisms into simpler compounds. All these products combinely influence the physical and chemical properties of soil and hence the plant growth.

Inorganic transformations. These processes are carried out by different soil micro-organisms. Through this process organic form of different nutrients like N, P, S, Fe etc. are transformed into their inorganic form and available to plants for their growth.

Fixation of nitrogen: Bacteria: symbiotic nitrogen fixing bacteria, e.g. Rhizobium. Non-symbiotic nitrogen fixing bacteria, e.g. Azotobacter, Azospirillum etc.

Cynobacteria: For example, Nostoc, Aulosira, Fertiliima, Anabaena etc.

Actinomycetes for example, Frankia.

Formation and development of soil.

Decomposition of rocks and minerals

Improvement of soil structure and stability of soil aggregates.

Movement of soil materials etc.

Production of soil enzymes: For example, hydrolysis of urea by urease enzyme.

Harmful: Denitrification process carried out by a particular micro-organisms and causes loss of gaseous nitrogen.

Development of various plant diseases.

Production of antibodies and other toxic substances.

Competition for nutrients.

Origin of organic matter: Soil organic matter mainly originates from the plant tissue. Leaves and roots of trees, shrubs, grasses and other plants etc. usually supply large quantities of organic materials to the soil. These plant constituents from the primary material both for the food of the soil organisms and for the production of soil organic matter. Animals are generally considered secondary sources of organic matter. As they break down the original plant tissue, they contribute the waste products and leave their own bodies after death. Certain forms of animal life, particularly earthworms, centipedes, insects and ants, also play an important role in the turnover of plant residues.

Simple outline for presence of different compounds in organic residues.

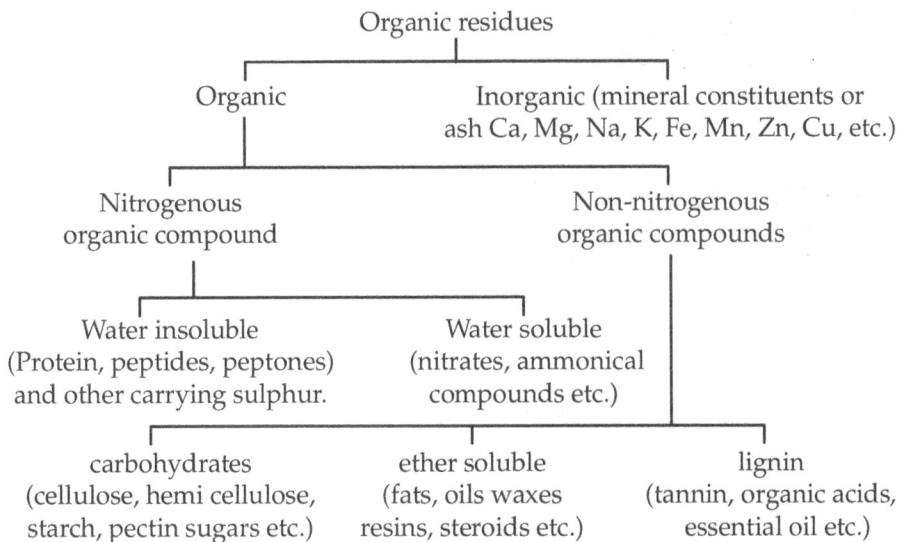

Organic residues

Organic — Inorganic (mineral constituents or ash Ca, Mg, Na, K, Fe, Mn, Zn, Cu, etc.)

Nitrogenous organic compound — Non-nitrogenous organic compounds

Water insoluble (Protein, peptides, peptones) and other carrying sulphur. — Water soluble (nitrates, ammonical compounds etc.)

carbohydrates (cellulose, hemi cellulose, starch, pectin sugars etc.) — ether soluble (fats, oils waxes resins, steroids etc.) — lignin (tannin, organic acids, essential oil etc.)

39. Common enzymes and their particular reactions

(a) **Urease:** Breaks down urea

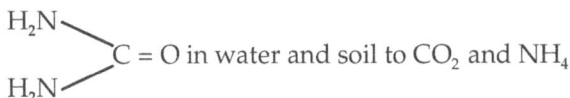

H_2N

$C = O$ in water and soil to CO_2 and NH_4

H_2N

(b) **Protease:** By involving water, it breaks the bond linking two amino acids to form separate amino acids or parts of proteins.

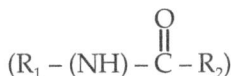

$$(R_1 - (NH) - \overset{\displaystyle O}{\overset{\displaystyle \|}{C}} - R_2)$$

$$(R_1 - NH_2 \text{ and } HO - \overset{\displaystyle O}{\overset{\|}{C}} - R_2)$$

(c) **Cellulose:** break celluloses (cell-wall fibers "wood") which are long chains of sugar units. Important in decomposition of organic matter.

(d) **Ligniase or lignase:** Degrades lignin compounds of decomposed organic residues.

(e) **Hemicellulose:** Breaks hemicelluloses compounds of organic materials.

(f) **Phosphatase:** by involving water it breaks

"humus $- \overset{\displaystyle O}{\overset{\|}{o}} - p - (OH)_2$" bond to produce "humus – OH" and H_3PO_4 (helps to decompose humus), making phosphorus available to the plant.

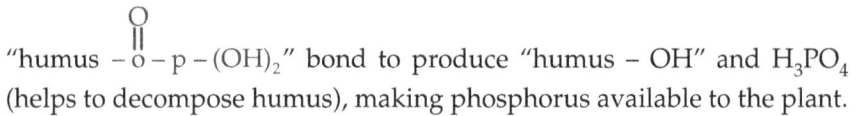

(g) **Sulphatase:** by involving water, it breaks the

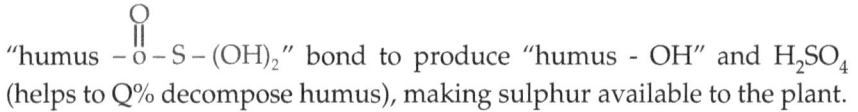

"humus $- \overset{\displaystyle O}{\overset{\|}{o}} - S - (OH)_2$" bond to produce "humus - OH" and H_2SO_4 (helps to Q% decompose humus), making sulphur available to the plant.

(h) **proto pectinase:** Decompose protopectin to form soluble pectin.

(i) **Pectin methylesterase (PME):** Hydrogen methyl ester lingages of pectin to form pectic acid and methanol.

(j) **Poly galacturonase (PG):** Breaks linkages between galacturonic acid and pectin compounds to form free galacturonic acid.

40. Carbon: Nitrogen (C:N) Ratio

It is the intimate relationship between organic matter and nitrogen content of soils. The ratio of the weight of organic carbon to the weight of total nitrogen in a soil or organic material is known as C:N ration. The C: N ratio of soil is one of its characteristics equilibrium values, the figure of humus being roughly 10:1 although values from 5:1 to 15:1 are generally found in most arable soils. The C:N ratio in plant material is variable and ranges from 20:1 to 30:1 to legumes and farm yard manure to as high as 10:1 in certain straw residues. The C:N ratio of micro-organisms is much narrower between 4:1 and 9:1.

41. Forms of nutrient elements absorbed by plants

(1) Absorb as a single nutrient element

Nutrient element	Forms
Potassium	K^+
Calcium	Ca^{2+}

Nutrient element	Forms
Magnesium	Mg^{2+}
Iron	Fe^{2+} (Ferrous), Fe^{3+} (Ferric)
Manganese	Mn^{2+}(Mangnous), Mn^{4+} (Manganic)
Copper	Cu^+ (cuprous), Cu^{2+} (cupric)
Zinc	Zn^{2+}
Chlorine	Cl^-
Silicon	$Si(OH)_4$
Cobalt	Co^{2+}
Sodium	Na^+

(2) *Absorb as a combined form :*

Nutrient element	Forms
Nitrogen	Ammonium $(NH_4{}^+)$, Nitrate $(NO_3{}^-)$
Phosphorus	$HPO_4{}^2$, $H_2Po_4{}^-$
Sulphur	$SO_4{}^{2-}$
Boron	H_3BO_3 (Boric acid), H_2BO_2 (Borate)
Molybdenum	$M_0O_4{}^{2-}$ (Molybdate)
Carbon	CO_2
Oxygen	CO_2, O^{2-}, OH^-, $CO_3{}^{2-}$
Hydrogen	HOH, H^+
Water	H^+, OH^-.

42. Classification of plant nutrients

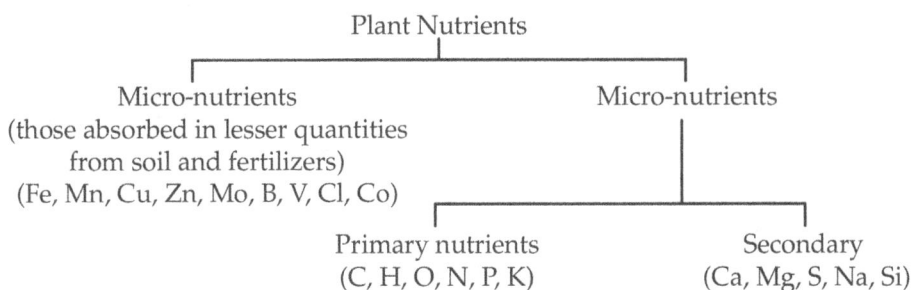

Plant Nutrients

Micro-nutrients
(those absorbed in lesser quantities
from soil and fertilizers)
(Fe, Mn, Cu, Zn, Mo, B, V, Cl, Co)

Micro-nutrients

Primary nutrients
(C, H, O, N, P, K)

Secondary
(Ca, Mg, S, Na, Si)

43. Mechanisms of nutrient uptake

Mass flow: Movement of ions in the soil solution to the surfaces of roots is an important factor in satisfying the nutrient requirement of plants. This movement is accomplished largely by mass flow and diffusion. Mass flow is a convective process, in which plant nutrient ions and other dissolved substances are transported in the flow of water to the root due to transpirational water uptake by the plant.

Diffusion: Diffusion process operates when an ion moves from an area of high concentration to one of low concentration by random thermal motion. As plant roots absorb nutrients from the surrounding soil solution, a diffusion gradient is set up. A high plant requirement or a high root absorbing power results in a strong sink or a high diffusion gradient favouring ion transport.

Root interception: The importance of root interception mechanism for ion absorption is enhanced by the growth of new roots throughout the soil mass and probably also by mycorrhizal infection. As the root system develops and exploits the soil more completely soil solution and soil surfaces retaining adsorbed ions are exposed to the root mass and absorption of these ions by the contact exchange phenomenon is accomplished.

44. Ion uptake mechanisms

The mechanism by which plants take up nutrients must be selective.

Selectivity: Certain mineral elements are taken up preferentially, which others are discriminated against or almost excluded.

Accumulation: The concentration of mineral nutrients can be much higher in the plant cell sap than in the external solution.

Genotype: There are distinct differences among plant species in ion uptake characteristics.

45. Chelates

The term "chelate" is derived from a Greek work which means "claw". Chelate is defined as a type of chemical compound in which a metallic ion is firmly combined with a molecule by means of multiple chemical bonds. The chelates have a marked tendency to hold tightly certain cations attached to them. The chelates may be formed either naturally between organic matter and metal ions (natural chelates) or artificially between different chelating agents [Diethylene triamine penta acetic acid (DTPA), EDTA (Ethylene diamine tetra acetic acid) etc. and metal ions (synthetic chelates).

Common chemical names of some chelating agents:

Names	Abbreviations
1. Citric acid	CIT
2. Diethylene triamine penta-acetic acid	DTPA
3. Ethylene diamine tetra acetic acid	EDTA
4. Ethylene diamine –N-N' diacetic acid	EDDA
5. N- (2-hydroxy ethyl) ethylene dinitrilo-triacetic acid	(HEEDTA)

46. Transformation of nitrogen in soils

Mineralisation: The process by which nitrogen in organic compound becomes converted into the inorganic ammonium (NH_4^+) and (NO_3^-) ions carried out by different soil micro-organisms.

$$\text{Organic N} \quad \underset{\text{(Amine)}}{R\text{-}NH_2} \longrightarrow \underset{\text{(Ammonium)}}{NH_4^+} \longrightarrow \underset{\text{(Nitrite)}}{NO_2^-} \longrightarrow \underset{\text{(Nitrite)}}{NO_2^-}$$

(a) **Aminization :** The process by which the hydrolytic decomposition of proteins from combined nitrogenous compounds as well as release of amines and amino acids takes place by heterotrophic soil micro-organisms.

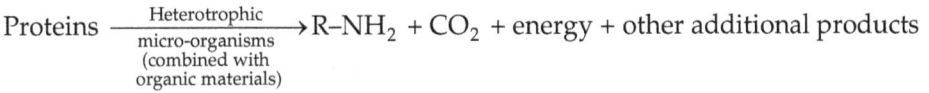

$$\text{Proteins} \xrightarrow[\substack{\text{micro-organisms} \\ \text{(combined with} \\ \text{organic materials)}}]{\text{Heterotrophic}} R\text{-}NH_2 + CO_2 + \text{energy} + \text{other additional products}$$

(b) **Ammonification :** The reduction of amines to ammonical compounds by heterotrophic micro-organisms is known as ammonification.

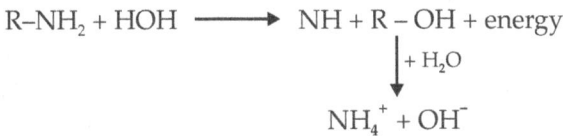

$$R\text{-}NH_2 + HOH \longrightarrow NH + R - OH + \text{energy}$$
$$\downarrow + H_2O$$
$$NH_4^+ + OH^-$$

(c) **Nitrification :** the process by which the microbial oxidation of ammonical nitrogen to nitrate form of nitrogen takes place.

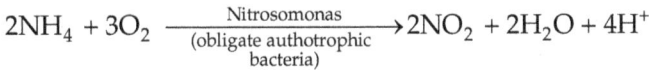

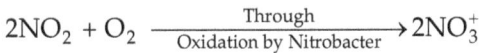

$$2NH_4 + 3O_2 \xrightarrow[\substack{\text{(obligate authotrophic} \\ \text{bacteria)}}]{\text{Nitrosomonas}} 2NO_2 + 2H_2O + 4H^+$$

$$2NO_2 + O_2 \xrightarrow[\text{Oxidation by Nitrobacter}]{\text{Through}} 2NO_3^+$$

Immobilization: The change of mineral nitrogen to organic forms by soil micro-organisms. Assimilation is sometimes used in the same sense as immobilization, whether immobilization will take place or not depends upon the initial nitrogen content in the organic materials applied to the soil.

47. Chemistry and behaviour of phosphorus in soils

Organic phosphorus : During mineralization of organic phosphorus substances, the release of inorganic phosphorus takes place in soil solution and such released phosphorus reacts very quickly with various soil components forming insoluble complex phosphatic compounds and there by unavailable to the plants. Mineralisation of phosphorus is of three types:

(a) Based on the lowering of organic phosphorus level in soil due to long term cultivation.

(b) Based on the results of short laboratory investigations decreasing the level of organic phosphorus with simultaneous increase in the amount of inorganic phosphorus in the soil.

(c) Based on monitoring levels of soil organic phosphorus in the presence and absence of plants considering seasonal variation.

(2) Inorganic phosphorus : It is evident that most of the soluble inorganic phosphorus either released from the mineralization of organic phosphorus or applied as soluble phosphatic fertilizers and rendered unavailable to the plants and hardly 20% of the applied phosphatic fertilizers are available to the plants. The reasons for such recovery are the conversions of soluble forms of phosphorus to a form which is very less soluble through reactions with various soil components involving different mechanisms. Such mechanisms for the removal of phosphorus from solution phase in the soil are known as "retention or fixation". However the retention of phosphorus in the soil involves various mechanisms namely, sorption and precipitation reactions.

48. Behaviour of potassium in soils

Soil solution potassium : It is recognized as the readily available form of potassium to the plants. The potassium availability in soils is controlled not only by the soil solution potassium but also by its buffering capacity (ability of a soil to maintain potassium intensity) the soil solution potassium (Intensity, I) is maintained by the exchangeable potassium in a dynamic equilibrium. Soil solution potassium content usually higher in arid region and saline soil ranging from 3 to 156 ppm whereas the content of the same of the same is lower in humid region soils ranging from 1 to 80 ppm concentration of water soluble potassium may be as low as 8 ppm in deficient soils.

Exchangeable potassium : Potassium ion (K^+) is held by soil colloids through electrostatic attraction similar to other cations. However, potassium held by soil colloids is easily displaced or exchanged when extracting the soil with neutral salt solutions. The amount of K exchanged varies with cations and usually neutral normal ammonium acetate solution is used for the purpose. A small amount of potassium in this fraction occurs in soils (< 1.0% of the total potassium). The distribution of potassium on soil colloids as well as soil solution depends upon nature and amounts of complementary cations, anion concentration and nature and characteristics of clay minerals.

Non-Exchangeable and mineral form of potassium

Potassium in these forms is not readily available to the plants. However, non-exchangeable potassium pools not instantly available to plants, can contribute significantly to the maintenance of the labile pool of potassium in the soil. On the other hand in some soils these fractions of potassium may become available as water soluble and exchangeable forms are removed by leaching, crop uptake etc. These forms of potassium are consisting of different K-bearing minerals namely primary minerals (K-feldspars) and micas (muscovites, biotites etc.), originating from the parent rock and secondary minerals (clays of the illitic group) formed by alteration of micas.

49. Properties of submerged soil

(A) *Physical properties*

1. Diffusion of molecular oxygen and development of aerobic-anaerobic layer: When a soil is submerged, water replaces the air in the pore spaces. Except in a thin layer at the soil surface, and sometimes a layer below the plough sole, most soil layers are virtually oxygen-free within a few hours after submergence. The oxygen diffusion in the water layer above the soil is very slow and the rate of O_2 consumption is reduced soil is high. Because of this high demand of O_2 in submerged soil and slow O_2 supply through water, the soil is practically devoid of oxygen. This rapid depletion of O_2 takes place within a day or so of submergence. Some O_2 trapped in blocked pore spaces is rapidly utilized by facultative anaerobic organisms.

2. Aeration status of submerged soil : Immediately after submergence, the normal process of gaseous exchange between soil and air is restricted. The magnitude of this effect is evident from the relative values in air and in water of the diffusion coefficient D, in the following equation.

$$V = -a\ D\ (T/T_0)^2 \cdot dp/dl$$

Where,

V = volume of gas,   dp/dl = Pressure gradient

a = porosity factor   T = absolute temperature

3. Accumulation of carbon dioxide : Soil gases like CO_2 and methane (CH_4) accumulate due to submergence and also may escape as bubbles if pressure builds up. It has been recorded that during the first three weeks submergenece some soils may generate CO_2 upto 2.5 t ha^{-1}.

4. Soil compaction : In compaction soil solids are rearranged with compression of liquid and gaseous phases accompanied by volume change. In compacted soil, bulk density, microvoids, thermal conductivity and diffusivity and nutrient mobility increases, and on the other hand, macrovoids hydraulic conductivity and water intake rates decreases. Generally medium textured soils are most susceptible to compaction.

5. Puddling : Puddling refers to breaking down soil aggregates at near saturation into ultimate soil particles,. The mechanical reduction in the apparent specific volume of soil is found due to puddling. Only soil with more than 20 per cent of clay particles is prone to puddling. Puddling influences physical, chemical and biological soil properties which in turn influence rice growth.

50. Classification of manures

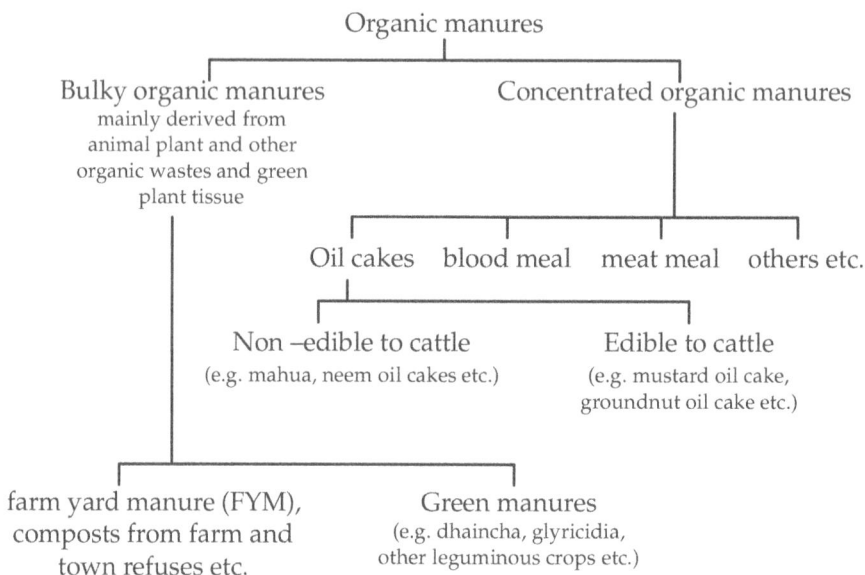

```
                          Organic manures
                               |
        ┌──────────────────────┴──────────────────────┐
  Bulky organic manures                Concentrated organic manures
    mainly derived from                              |
    animal plant and other                           |
    organic wastes and green                         |
    plant tissue                                     |
        |                  ┌──────────┬──────────┬──────────┐
        |              Oil cakes  blood meal  meat meal  others etc.
        |                  |
        |          ┌───────┴────────┐
        |     Non –edible to cattle      Edible to cattle
        |     (e.g. mahua, neem oil      (e.g. mustard oil cake,
        |      cakes etc.)                groundnut oil cake etc.)
        |
   ┌────┴──────────────────────────┐
farm yard manure (FYM),        Green manures
composts from farm and     (e.g. dhaincha, glyricidia,
town refuses etc.          other leguminous crops etc.)
```

51. Characteristics of organic manures

All the manures are bulky in nature as well as concentrated nature and supply: Plant nutrients in small quantities and organic matter in large amounts. When it is applied into the soil it will act as follows:

Organic manures supply primary, secondary and micro-nutrients to plants which are librated in an available forms during the process of mineralization carried out by different micro-organisms. Organic manures also supply organic matter to the soil and hence improved the physical condition of the soil like soil structure, aeration, water holding capacity etc. It also stimulates the activity of different soil micro-organisms through the supply of energy. It improves the buffering and exchange capacities of soil and also influences the solubility of soil minerals as well as mineral nutrients in soil. It also forms chelates which also help for the nutrition of plants. It also regulates the thermal regimes of the soil.

52. Fertilizers

Complete fertilizer : Complete fertilizer is referred to a fertilizer material which contains all three major nutrients N, P and K.

Incomplete fertilizer : This fertilizer is referred to a fertilizer material which lacks any one of three major nutrient elements.

Straight fertilizer : Straight fertilizers may be defined as chemical fertilizers which contain only one primary or major nutrient element, e.g., ammonium sulphate [$(NH_4)_2 SO_4$], urea [$CO(NH_2)_2$].

Mixed or complex fertilizer: These fertilizers may be defined as a fertilizer material which contains more than one primary or major nutrient elements produced by the process of chemical reactions.

Nitrogenous fertilizers : Nitrogenous fertilizers are those fertilizers that are sold for their nitrogen content. Nitrogenous fertilizers can be classified into four classes on the basis of forms of N present in straight nitrogenous fertilizers as follows:

(1) Nitrate nitrogen (NO_3 - N) containing fertilizers.

       e.g., $NaNo_3$ $\longrightarrow$ 16% N

                $Ca(NO_3)_2$ $\longrightarrow$ 15.5% N

(2) Ammonium containing nitrogenous fertilizers (NH_4-N)

       e.g., $(NH_4)_2SO_4$ $\longrightarrow$ 20% N

                NH_4Cl $\longrightarrow$ 24-26% N

       Anhydrous ammonia $\longrightarrow$ 82% N

(3) Both NH_4^+ and NO_3 – N containing nitrogenous fertilizers

       e.g., ammonium nitrate : (NH_4NO_3) $\longrightarrow$ 33 –34% N

       Calcium ammonium nitrate (CAN) $\longrightarrow$ 20% N

(4) Amide fertilizers: It is organic form of N-containing fertilizers

       e.g.,   Urea [$CO(NH_2)_2$] $\longrightarrow$ 46% N

       Calcium cyanamide ($CaCN_2$) $\longrightarrow$ 21% N

53. Slow release nitrogenous fertilizers

Slow release nitrogenous fertilizers minimize the loss of N increases the efficiency of the fertilizer.

It helps to avoid frequent application of N-fertilizers.

It helps to prevent different harmful effects of plants like germination of seeds and emergence of seedlings.

It avoids the luxury consumption of N and prevents the imbalances due to different nutrients within the plants.

Classification of slow release N- fertilizers: coated N-fertilizers, e.g., sulphur coated urea (SCU), Neem Coated Urea (NCU), Lac Coated Urea (LCU).

N-substances of low water solubility, e.g., Urea-formaldehyde or Urea form

Nitrification and urease inhibitors, e.g., N-serve. It is 2-chloro–6 (trichloromethyl) pyridine and also referred to as nitrapyrin.

AM – Chemically it is a substituted pyrimidine (2-amino-4-chloro-6-methylpyrimidine).

Sparingly soluble minerals.

54. Slow release nitrogenous fertilizers

Slow release nitrogenous fertilizers minimize the loss of N increases the efficiency of the fertilizer.

It helps to avoid frequent application of N-fertilizers.

It helps to prevent different harmful effects of plants like germination of seeds and emergence.

It avoids the luxury consumption of N and prevents the imbalances due to different nutrients within the plants.

Classification of slow release N–fertilizers:

1. Coated N-fertilizers

    e.g., Sulphur coated Urea (SCU)

    Neem Coated Urea (NCU)

    Lac Coated Urea (LCU)

2. N–substances of low water solubility.

    e.g., Urea–formaldehyde or Urea-form

3. Nitrification and urease inhibitors.

    e.g., N-serve–It is 2-chloro–6 (trichloromethyl) pyridine and also referred to as nitrapyrin.

    AM-Chemically it is substituted pyrimidine (2- amino-4-chloro-6-mehtylpyrimidine).

4. Sparingly soluble minerals.

55. Phosphatic Fertilizers

Phosphatic fertilizers can be classified into three groups on the basis of forms in which orthophosphoric acid (H_3Po_4) is combined with calcium (Ca^{2+}).

1. Water soluble monocalcium phosphate, $[Ca(H_2PO_4)_2]$

   e.g., superphosphates (SSP, DSP and TSP)

   Single Super Phosphate (SSP) : 16 – 18% P_2O_5 or 6.88 – 7.74% P

   Double Super Phosphate (DSP) : 32% P_2O_5 or 13.76% P

   Triple Super Phosphate (TSP) : 46 – 48% P_2O_5 or 19.78 – 20.64% P

2. Citric acid soluble, dicalcium phosphate : $[Ca_3\,2H_2\,(PO_4)_2,$ or $CaHPO_4]$

   e.g., Basic slag, silicates of lime : 14-18% P_2O_5 or 6.02 – 7.74% P

   Dicalcium phosphate : 34-39% P_2O_5 or 14.62 – 16.77% P

3. Phosphatic fertilizers not soluble in water or not soluble in citric acid, tricalcium phosphate, $[Ca_3\,(PO_4)_2]$.

   e.g., Rock phosphate : 20-40% P_2O_5 or 8.6 – 10.75% P

   Steamed bonemeal : 22% P_2O_5 or 9.46 % P

56. Types of Erosion

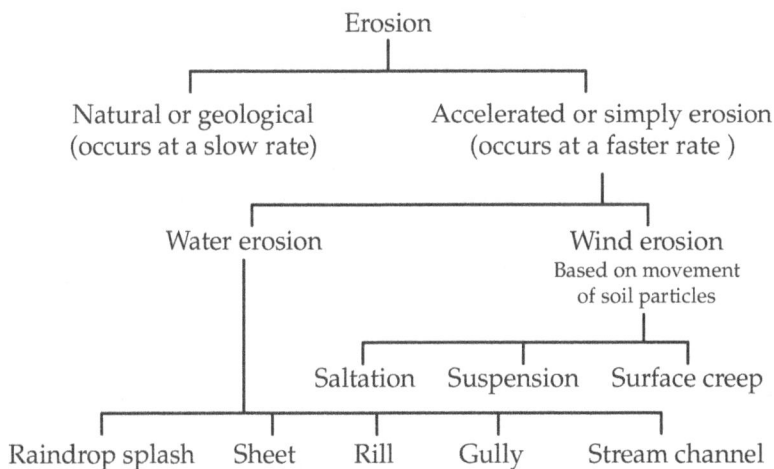

57. Causes of water erosion

Raindrop splash erosion: Raindrop splash erosion results from soil splash caused by the impact of falling raindrops the impact of raindrops per unit area is determined by the number and size of the drops, and the velocity of the drops.

Sheet erosion: Sheet erosion is the removal of a fairly uniform layer of surface soil by the action of rainfall and runoff water. This type of erosion, through extremely harmful to the land. It is common on lands having a gentle or mild slope, and results in the uniform "skimming of the cream" of the top soil with every hard rain. Movement of soil by rain drop splash is the primary cause of sheet erosion.

Rill erosion : Rill erosion is the removal of surface soil by running water, with the formation of narrow shallow channels that can be leveled or smoothed out completely by normal cultivation. Rill erosion is more apparent than sheet erosion. This type of soil erosion may be regarded as a transition state between sheet and gully erosion.

Gully erosion : Gully erosion is the removal of soil by running water, with the formation of channels that cannot be smoothed out completely by normal agricultural operation or cultivation. Gully erosion is an advanced stage of rill erosion. Unattended rills get deepened and widened every year and begin to attain the form of gullies. Gully erosion is more spectacular, and therefore, more noticeable than any other erosion.

Stream channel erosion : Stream channel erosion is the scouring of material from the water running water. This erosion occurs at the lower end of stream tributaries and to streams that have nearly continuous flow and relatively flat gradients.

58. Wind erodes the soil in three steps

Saltation : Saltation is a process of soil movement in a series of bounces or jumps. Soil particles having sizes ranging from 0.05 to 0.5 mm generally move in this process. Saltation movement is caused by the pressure of the wind on the soil particle, and collision of a particle with other particles. The height of the jumps varies with the size and density to the soil particles, the roughness of the soil surface, and the velocity of the wind.

Suspension : Suspension represents the floating of small sized particles in the air stream. Movement of such fine particles in suspension is usually started by the impact of particles in saltation. Once these fine particles are picked up by the particles in saltation and enter the turbulent air layers, they can be lifted upward in the air by eddy current and they are often carried for several miles before being redeposited elsewhere.

Surface creep : Surface creep is the rolling or sliding of large soil particles along the ground surface. They are too heavy to be lifted by the wind and are moved primarily by the impact of the particles in siltation rather than by direct force of the wind. The coarse particles tend to move closer to the ground than the fine ones.

CLASSIFICATION

1. *Volume composition of soil*

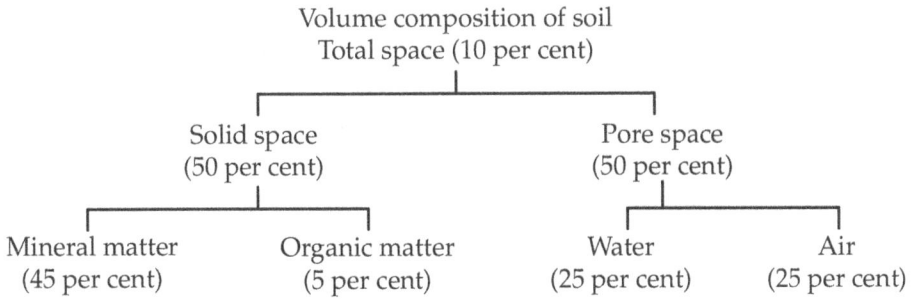

Volume composition of soil
Total space (10 per cent)

Solid space (50 per cent) — Pore space (50 per cent)

Mineral matter (45 per cent) — Organic matter (5 per cent) — Water (25 per cent) — Air (25 per cent)

2. *Classification of soil forming rocks*

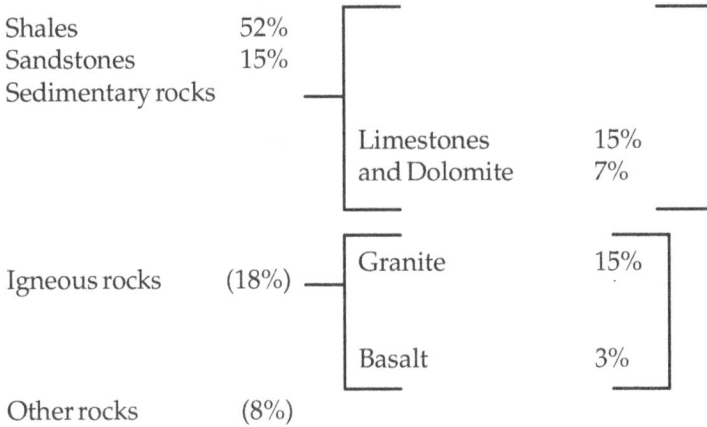

Shales	52%	
Sandstones	15%	Sedimentary rocks
Sedimentary rocks		
	Limestones	15%
	and Dolomite	7%
Igneous rocks	(18%)	Granite 15%
		Basalt 3%
Other rocks	(8%)	

Igneous Rocks

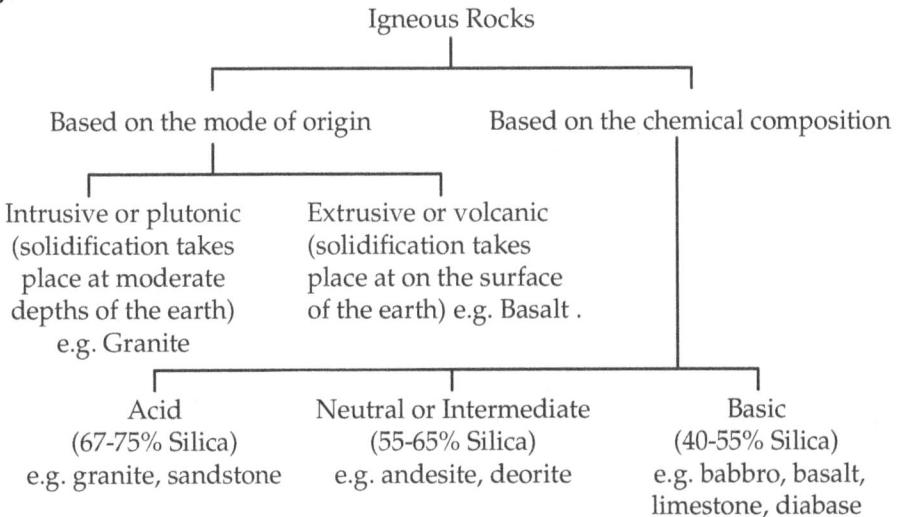

Igneous Rocks

Based on the mode of origin — Based on the chemical composition

Intrusive or plutonic (solidification takes place at moderate depths of the earth) e.g. Granite

Extrusive or volcanic (solidification takes place at on the surface of the earth) e.g. Basalt .

Acid (67-75% Silica) e.g. granite, sandstone

Neutral or Intermediate (55-65% Silica) e.g. andesite, deorite

Basic (40-55% Silica) e.g. babbro, basalt, limestone, diabase

Sedimentary Rocks

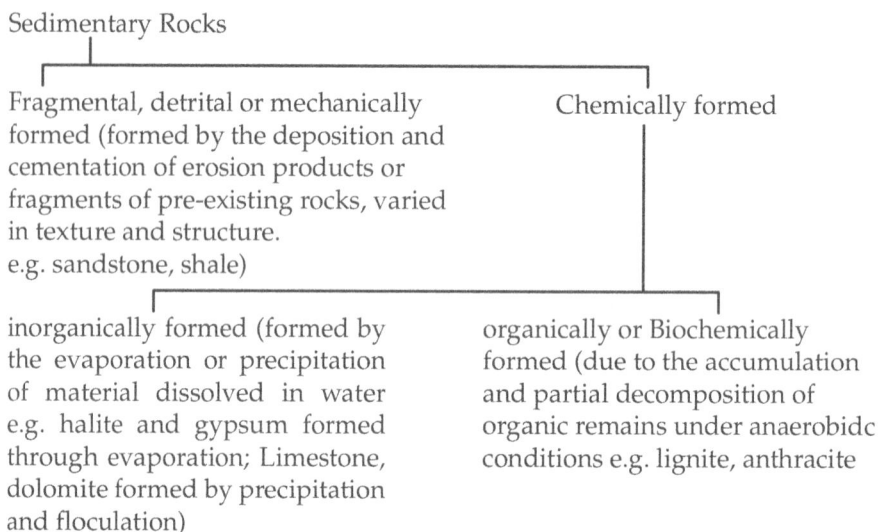

Sedimentary Rocks

Fragmental, detrital or mechanically formed (formed by the deposition and cementation of erosion products or fragments of pre-existing rocks, varied in texture and structure. e.g. sandstone, shale)

Chemically formed

inorganically formed (formed by the evaporation or precipitation of material dissolved in water e.g. halite and gypsum formed through evaporation; Limestone, dolomite formed by precipitation and floculation)

organically or Biochemically formed (due to the accumulation and partial decomposition of organic remains under anaerobidc conditions e.g. lignite, anthracite)

3. Classification of Minerals based on different parameters

(A) Based on amount

Essential minerals : Form major parts of rocks. Present in quantities varying from 95-98%, e.g., calcite, silicate minerals.

Accessory minerals: Variety of minerals occurs in very small but significant quantities. (2-5% e.g. appetite, pyrite, magnetite).

(B) Based on mode of origin

Primary : Inherited from the parent rocks, make up the main part of the sand and silt fractions of soil and formed of elevated temperature, e.g., micas, hornblende, and quartz.

Secondary : Formed by low temperature reactions and inherited by soils from sedimentary rocks or formed in soils by weathering, e.g., serpentine, clay minerals.

(C) Based on specific gravity

Light minerals: Having specific gravity < 2.85, e.g., quartz (2.6), feldspar (2.65) muscovite (2.5 to 2.75).

Heavy minerals: Having specific gravity > 2.85 e.g. hematite (5.3), pyrite (5.0), limonite (3.8), augite (3.1 – 3.6), hornblende, amphiboles (2.9 – 9.8), ovivine (3.5).

(D) Based on Chemical Composition

Native elements : For example, graphite, sulphur, gold, copper etc.

Oxides and hydroxides : For example, quartz (SiO_2), hematite (Fe_2O_3).

Sulphur bearing minerals :

Sulphate bearings : For example, gypsum ($CaSO_4 . 2H_2O$)

Sulphide bearings : For example, pyrite ($FeS_2.CuS_2$)

Carbonate bearing minerals : For example, calcite ($CaCO_3$)

Halides : For example, rock salt (NaCl)

Silicates : For example, orthoclase, micas, olivine etc.

4. Classification of parent materials from which soils are derived

(A) Residual (remained in place for a long time).

Igneous : Granite, Basalt, Rhyolite

Sedimentary : Limestone, sandstone, shale

Metamorphic :

Igneous Pressure	Heat Genesis	Schist
Sedimentary Chemical	Quartzite	Marble

4. Classification of parent materials from which soils are derived

(A) Residual (remained in place for a long time)

1. **Igneous** : granite, basalt, Rhyolite.

2. **Sedimentary** : Limestone, sandstone, shade.

3. **Metamorphic**

   Igneous : Heat schist

   Pressure : Gnesis

   Sedimentary chemical Quartzite Marble

(B) Transported (movement of minerals and rock fragments)

1. **Water** : Marine deposit

   Lacustrine deposit

   Alluvial deposit

   Fanglomerate (cemented alluvial fan)

2. **Wind :** Eolians and sizes

   Loess (silt - size mostly)

3. **Ice (Glacial) :** Moraine : lateral or terminal ground moraine, outwash plain.

4. **Gravity :** Colluvial (Sides, mud flows)

   Organic accumulation or cumulose materials

   Moss peats

   Shrub and tree peats

   Mixed mucks

5. *Classification of soil particles*

(A) British standards institution

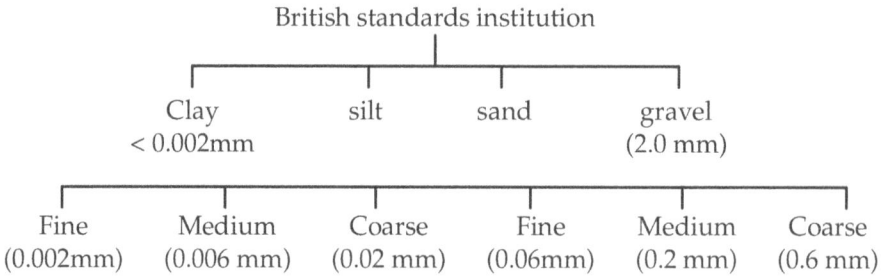

British standards institution

Clay	silt	sand	gravel
< 0.002mm			(2.0 mm)

Fine	Medium	Coarse	Fine	Medium	Coarse
(0.002mm)	(0.006 mm)	(0.02 mm)	(0.06mm)	(0.2 mm)	(0.6 mm)

(B) International union of soil science

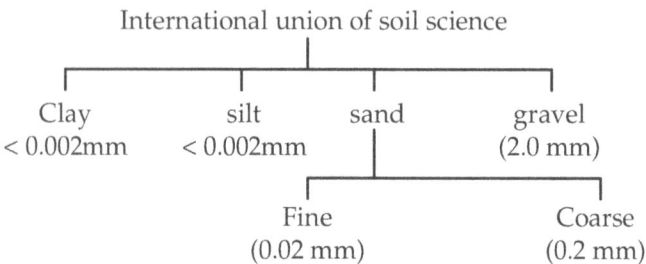

International union of soil science

International union of soil science

Clay	silt	sand	gravel
< 0.002mm	< 0.002mm		(2.0 mm)

Fine	Coarse
(0.02 mm)	(0.2 mm)

(C) United states department of Agriculture

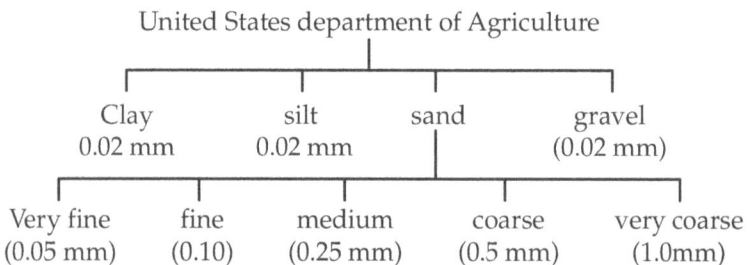

United States department of Agriculture

Clay	silt	sand	gravel
0.02 mm	0.02 mm		(0.02 mm)

Very fine	fine	medium	coarse	very coarse
(0.05 mm)	(0.10)	(0.25 mm)	(0.5 mm)	(1.0mm)

(D) United States Public Roads Administration:

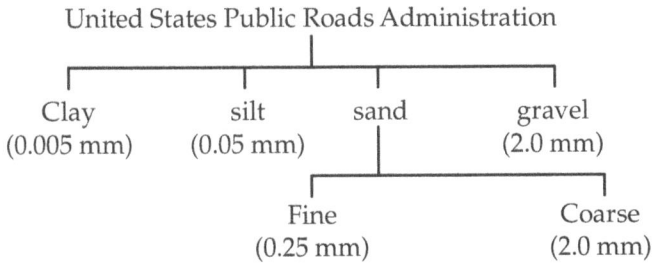

6. *Soil Textural Classes Names Developed by U.S. Department of Agriculture*

Common name	texture basic soil		Textural class name
(A) Sandy soils	coarse	⟶	sandy
Loamy sands			
(B) Loamy soils	Moderately	⟶	Sandy loam
Coarse	Fine sandy loam	⟶	
Medium	Loam		
Silt loam			
Silt			
	Moderately clay	⟶	Loam
Fine	Sandy clay loam		
Silty clay loam			
(C) Clavey soils	Fine	⟶	Sandy clay
Silty clay			
Clay			

7. *Classification of soil structure*

(A) Types of structure:

1. Plate like ⟶ laminar

2. Prism like

3. Block Like:

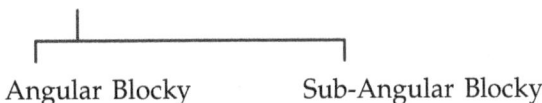

4. Spherical (Sphere like)

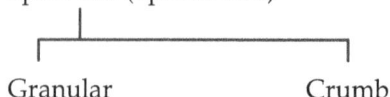

```
                |
    ┌───────────┴───────────────┐
 Granular                    Crumb
```

(B) Classes of soil structure

1. Very fine or very thin (in case of platy soil structure).
2. Fine or thin
3. Medium
4. Coarse or thick (in case of platy soil structure)
5. Very coarse or very thick

(C) Grades of soil structure

1. Structureless
2. Weak
3. Moderate
4. Strong

8. Composition of Soil and Atmospheric Air

Name of gas	Percentage by volume	
	Soil air	Atmospheric air
Oxygen	20.00	21.00
Nitrogen	78.60	78.03
Carbon dioxide	0.50	0.03
Argon	0.90	0.94

9. Classification of soil water

(A) Physical classification

Gravitational water: Held at a potential greater than −1/3 bar.

Capillary water: Retained in the soil between the water potential of −1/3 bar to −31 bars.

Hygroscopic water: Held by the soil particles of a solution more than −31 bars.

(B) Biological classification

Available water: Retained in the soil between field capacity (–1/3 bar) and the permanent wilting coefficients (-15 bars).

Unavailable water: Held at soil water potential greater than –15 bars.

Superfluous water: Retained in the soil beyond the field capacity soil moisture tension.

10. Types of soil colloids

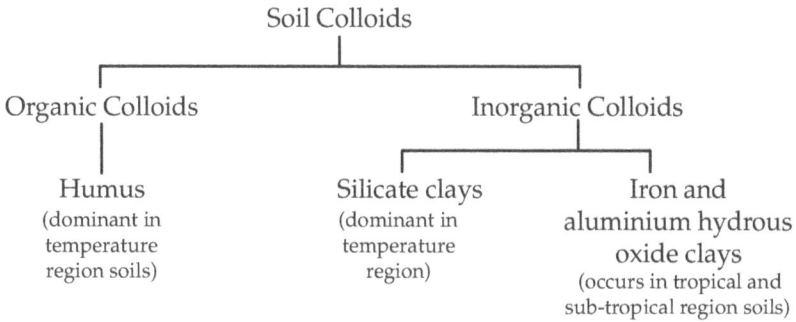

Soil Colloids

Organic Colloids   Inorganic Colloids

Humus
(dominant in temperature region soils)

Silicate clays
(dominant in temperature region)

Iron and aluminium hydrous oxide clays
(occurs in tropical and sub-tropical region soils)

11. Classification of silicate clays

On the basis of the number and arrangement of silica and alumina sheets, silicate clays may be classified into four different groups.

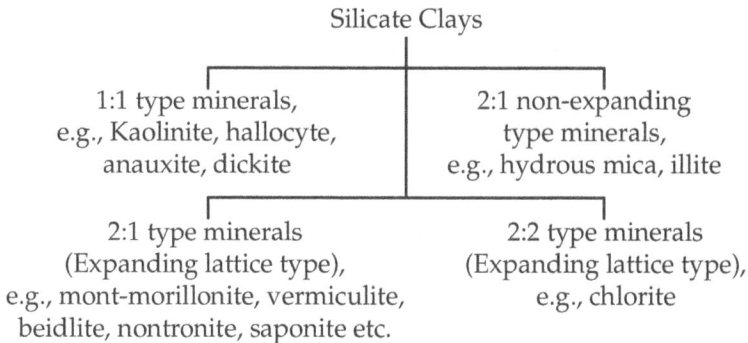

Silicate Clays

1:1 type minerals, e.g., Kaolinite, hallocyte, anauxite, dickite

2:1 non-expanding type minerals, e.g., hydrous mica, illite

2:1 type minerals (Expanding lattice type), e.g., mont-morillonite, vermiculite, beidlite, nontronite, saponite etc.

2:2 type minerals (Expanding lattice type), e.g., chlorite

12. Early systems of soil classification

Economic classification : Grouping soils according to their productivity was for purpose of taxation.

Physical classification : Based on soil textures.

Chemical classification: Based on chemical composition has not been used to a great extent in practical purposes.

Geological classification : Based on presumed underlying parent material.

Residual or sedentary soils: Developed in situ from the underlying rocks.

Transported soils: Developed on unconsolidated sediments, like alluvium, colluvium or Aeolian material.

Physiographic classification : The characteristics of the landscape were considered and so various geomorphic terms.

Other systems : Based on organic matter as:

Based on organic matter as:

Inorganic or mineral soils, and

Organic soils

Based on soil structure as:

Single grained soils, and

Aggregated soils

13. Soil classification by Dockuchaiev

Zonality concept developed by Russian soil scientist Dockuchaiev.

Zonal Soils : Soils developed under similar climate conditions and distributed in a climatic belt, e.g., laterite soils, podzol soil, chernozem soil etc.

Intrazonal soils : Soils occurs within a zone but reflect the influence of some local conditions, such as topography and the parent material, e.g., saline soil, saline sodic soil.

Azonal soils: soils have poorly developed profiles because of time as a limiting factor, e.g., Alluvial soils.

14. Name of soil orders in new comprehensive soil classification (7th Approximation)

1.	Entisol	7.	Alfisol
2.	Vertisol	8.	Ultisol
3.	Inceptisol	9.	Oxisol
4.	Aridisol	10.	Histosol
5.	Mollisol	11.	Andisol
6.	Spodosol	12.	Gelisol

15. *Land Capability classification system*

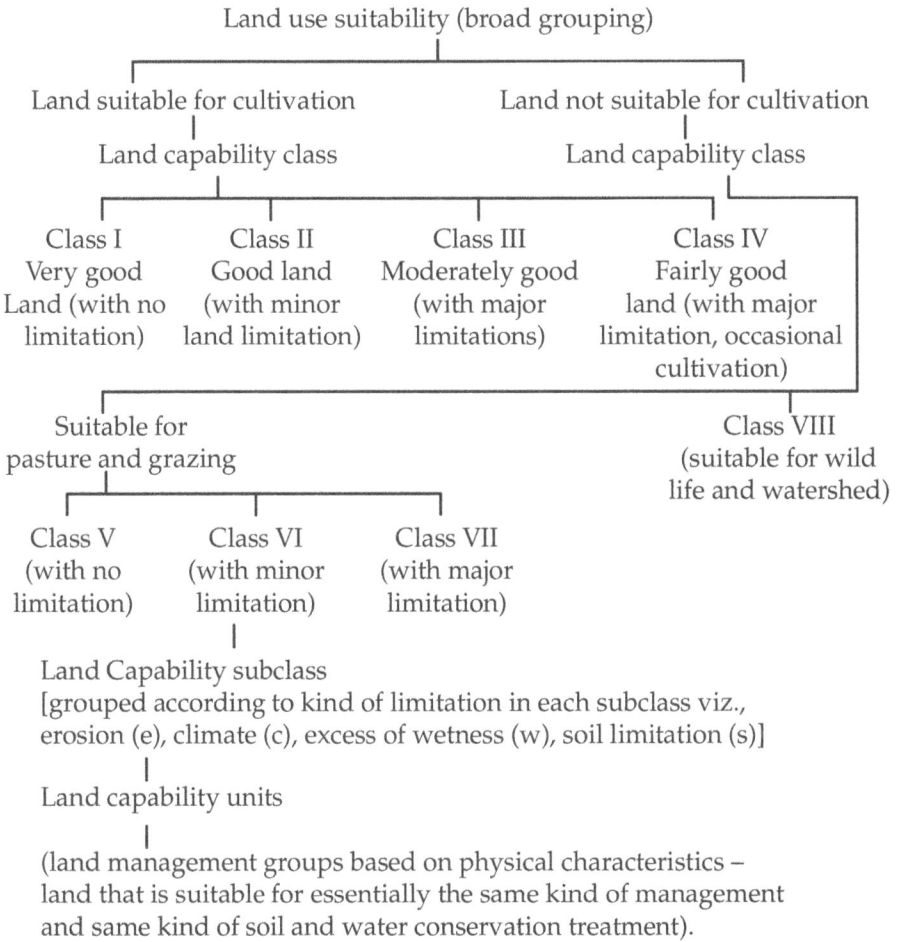

Land use suitability (broad grouping)

Land suitable for cultivation	Land not suitable for cultivation
Land capability class	Land capability class

Class I	Class II	Class III	Class IV
Very good Land (with no limitation)	Good land (with minor land limitation)	Moderately good (with major limitations)	Fairly good land (with major limitation, occasional cultivation)

Suitable for pasture and grazing

Class VIII (suitable for wild life and watershed)

Class V	Class VI	Class VII
(with no limitation)	(with minor limitation)	(with major limitation)

Land Capability subclass
[grouped according to kind of limitation in each subclass viz., erosion (e), climate (c), excess of wetness (w), soil limitation (s)]

Land capability units

(land management groups based on physical characteristics – land that is suitable for essentially the same kind of management and same kind of soil and water conservation treatment).

16. *Classification of acid soils in India*

Laterite

Laterite and lateritic red

Mixed red, black and yellow

Ferruginous red

Podsolic brown forest and forest soil

Foot hill soils

Peat soils

17. Classification of acid soils statewise with area and approximate pH range

Soil Groups	P^H range	Area (million hectares)	States
1. 3 laterites (plantha-guults, plitthu-stults, plinthudults oxisols)	4.8–7.0	12.65	Karnataka, M.P., West Bengal, South Maharashtra, Kerala, Malbar Coast, Assam, Bihar, Jharkhand
(i) Laterite in high rainfall zone with strongly expressed dry season.			
(ii) Laterite in high rainfall zone with weakly dry season.			
(iii) Laterite in sub-humid			
2. Laterite and lateritic red soils (plinthaquults, plinthudults oxisols, plinthustults)	5.0-7.0	11.80	Kerala, Orissa, West Bengal, Assam, M.P., Karnataka, Bihar, U.P. and Jharkhand
3. Mixed red and black/yellow soils (association of Alfisols and vertisols)	5.5-6.5	23.66	Karnataka, Bihar, M.P., U.P. and Jharkhand

18. Classification of soil organisms

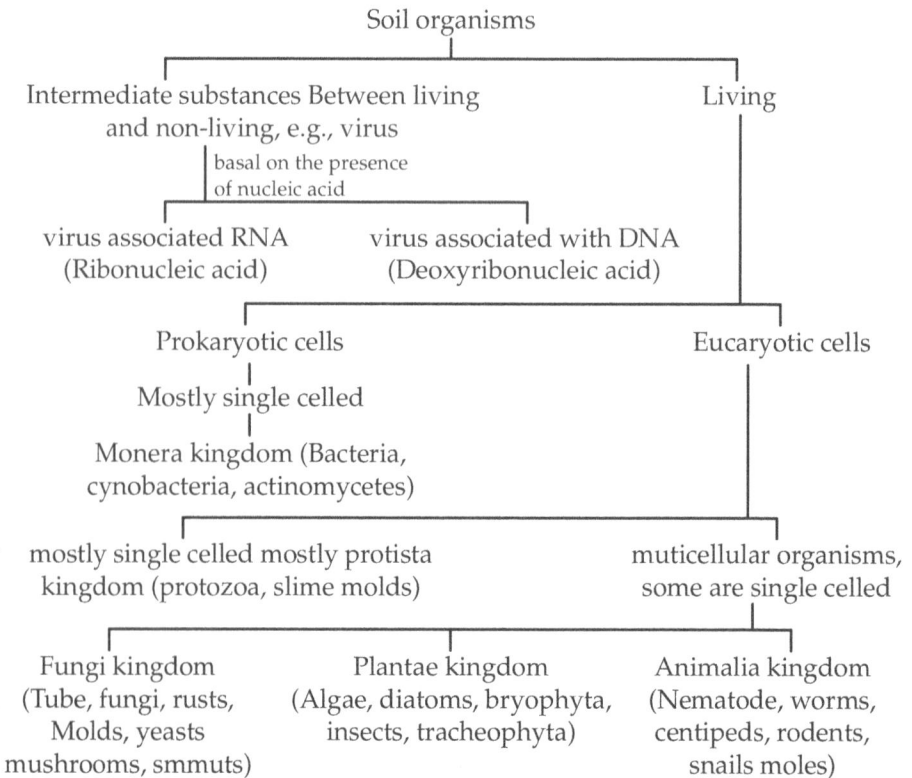

Soil organisms

Intermediate substances Between living and non-living, e.g., virus — Living

basal on the presence of nucleic acid

virus associated RNA (Ribonucleic acid) — virus associated with DNA (Deoxyribonucleic acid)

Prokaryotic cells — Eucaryotic cells

Mostly single celled

Monera kingdom (Bacteria, cynobacteria, actinomycetes)

mostly single celled mostly protista kingdom (protozoa, slime molds) — muticellular organisms, some are single celled

Fungi kingdom (Tube, fungi, rusts, Molds, yeasts mushrooms, smmuts) — Plantae kingdom (Algae, diatoms, bryophyta, insects, tracheophyta) — Animalia kingdom (Nematode, worms, centipeds, rodents, snails moles)

19. Classification of protozoa

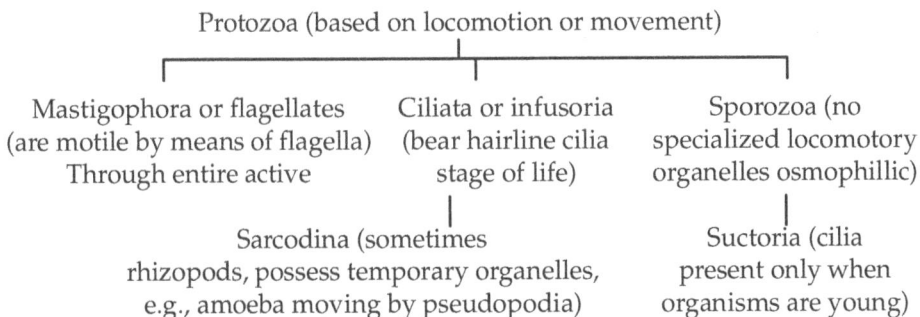

Protozoa (based on locomotion or movement)

Mastigophora or flagellates (are motile by means of flagella) Through entire active

Ciliata or infusoria (bear hairline cilia stage of life)

Sporozoa (no specialized locomotory organelles osmophillic)

Sarcodina (sometimes rhizopods, possess temporary organelles, e.g., amoeba moving by pseudopodia)

Suctoria (cilia present only when organisms are young)

20. Classification of Bacteria

(A) Based on nutrition and energy

1. Photoautotrophs (energy from sunlight, nutritive carbon from CO_2).

2. Photoautotrophs (energy from sunlight, carbon from organic matter)

3. Chemoautotrophs (energy from oxidation of inorganic substances such as N, Fe, or S, carbon from CO_2).

4. Chemoheterotrophs (energy and nutritive carbon from organic matter).

(B) Based on symbiotic relationship : Symbiotic dinitrogen (N_2) fixers (associated with a host plant, both host and bacteria benefit. Fixes N_2 from atmosphere).

Symbiotic dinitrogen (N_2) fixers (associated with a host plant, both host and bacteria benefit. Fixes N_2 from atmosphere).

21. Composition of organic residues

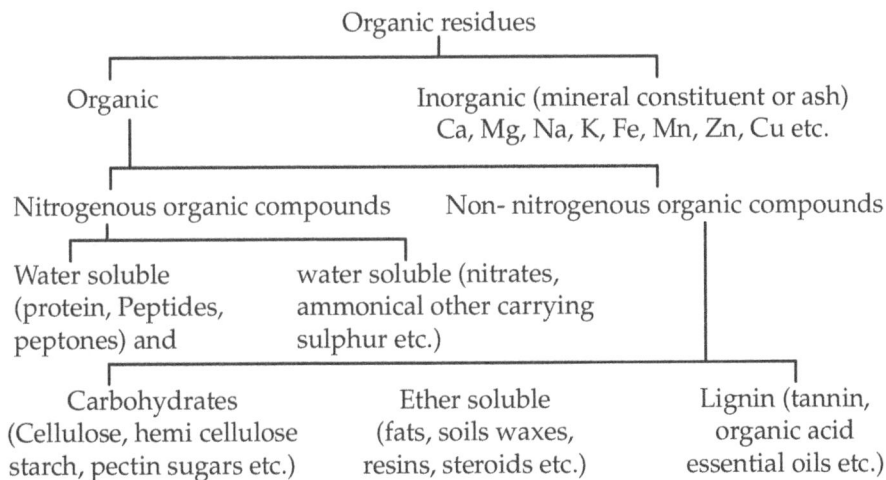

Organic residues

Organic

Inorganic (mineral constituent or ash) Ca, Mg, Na, K, Fe, Mn, Zn, Cu etc.

Nitrogenous organic compounds

Non- nitrogenous organic compounds

Water soluble (protein, Peptides, peptones) and

water soluble (nitrates, ammonical other carrying sulphur etc.)

Carbohydrates (Cellulose, hemi cellulose starch, pectin sugars etc.)

Ether soluble (fats, soils waxes, resins, steroids etc.)

Lignin (tannin, organic acid essential oils etc.)

22. Composition of a green plant material

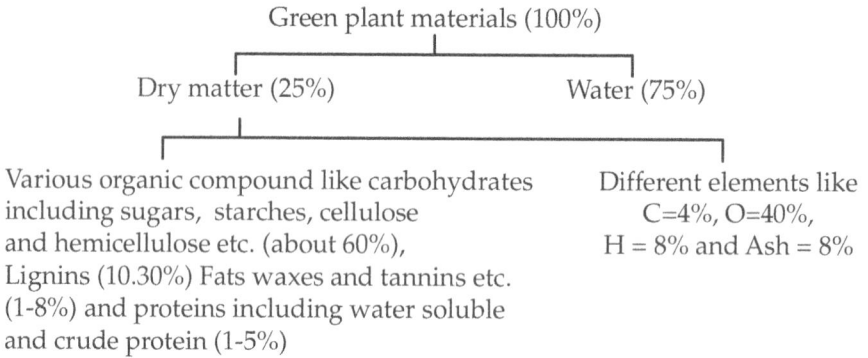

Green plant materials (100%)

Dry matter (25%)   Water (75%)

Various organic compound like carbohydrates including sugars, starches, cellulose and hemicellulose etc. (about 60%), Lignins (10.30%) Fats waxes and tannins etc. (1-8%) and proteins including water soluble and crude protein (1-5%)

Different elements like C=4%, O=40%, H = 8% and Ash = 8%

23. Classification of plant nutrients

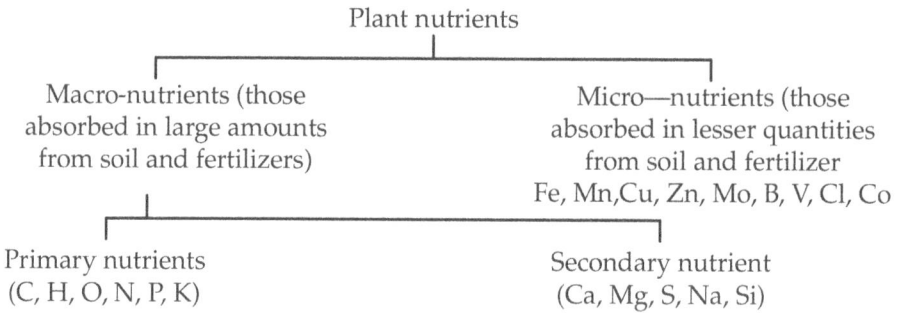

Plant nutrients

Macro-nutrients (those absorbed in large amounts from soil and fertilizers)

Micro—nutrients (those absorbed in lesser quantities from soil and fertilizer Fe, Mn,Cu, Zn, Mo, B, V, Cl, Co

Primary nutrients (C, H, O, N, P, K)

Secondary nutrient (Ca, Mg, S, Na, Si)

24. Classification of manures

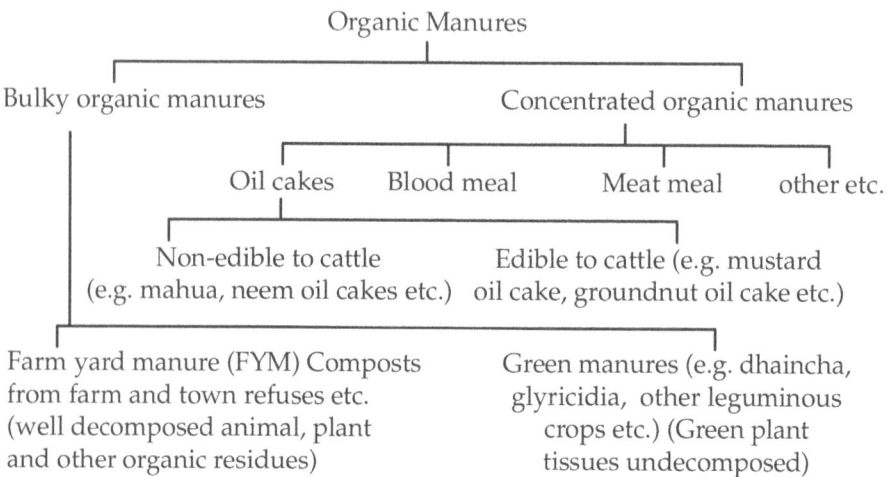

Organic Manures

Bulky organic manures

Concentrated organic manures

Oil cakes   Blood meal   Meat meal   other etc.

Non-edible to cattle (e.g. mahua, neem oil cakes etc.)

Edible to cattle (e.g. mustard oil cake, groundnut oil cake etc.)

Farm yard manure (FYM) Composts from farm and town refuses etc. (well decomposed animal, plant and other organic residues)

Green manures (e.g. dhaincha, glyricidia, other leguminous crops etc.) (Green plant tissues undecomposed)

25. *Kinds of green manureing*

Green manuring in situ: A system by which green manure crops are grown and incorporated into the soil of the same field that is to be green manured, either as a pure crop or an intercrop with either main crop.

e.g. sunhemp, dhaincha, guar etc.

Green manureing through collection of green plant tissues from other places. It refers to turning into the soil green leaves and tender green twing collected from outside the field to be green manured.

e.g. Glyricidia, Karanja etc.

26. *Classification of fertilizers*

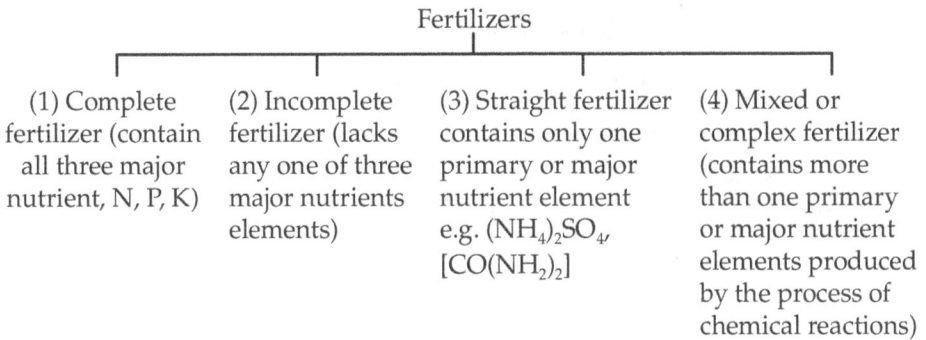

Fertilizers

(1) Complete fertilizer (contain all three major nutrient, N, P, K)	(2) Incomplete fertilizer (lacks any one of three major nutrients elements)	(3) Straight fertilizer contains only one primary or major nutrient element e.g. $(NH_4)_2SO_4$, $[CO(NH_2)_2]$	(4) Mixed or complex fertilizer (contains more than one primary or major nutrient elements produced by the process of chemical reactions)

27. *Classification of nitrogenous fertilizers*

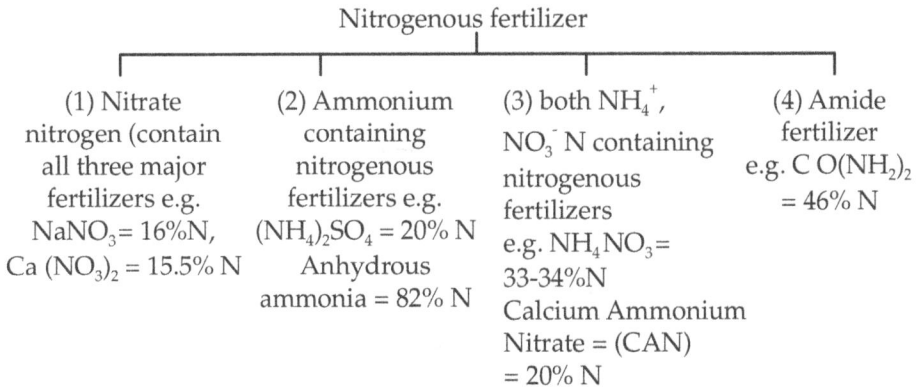

Nitrogenous fertilizer

(1) Nitrate nitrogen (contain all three major fertilizers e.g. $NaNO_3 = 16\%N$, $Ca(NO_3)_2 = 15.5\%\ N$	(2) Ammonium containing nitrogenous fertilizers e.g. $(NH_4)_2SO_4 = 20\%\ N$ Anhydrous ammonia = 82% N	(3) both NH_4^+, NO_3^- N containing nitrogenous fertilizers e.g. $NH_4NO_3 = 33\text{-}34\%N$ Calcium Ammonium Nitrate = (CAN) = 20% N	(4) Amide fertilizer e.g. $CO(NH_2)_2$ = 46% N

28. *Classification of slow release N-fertilizers*

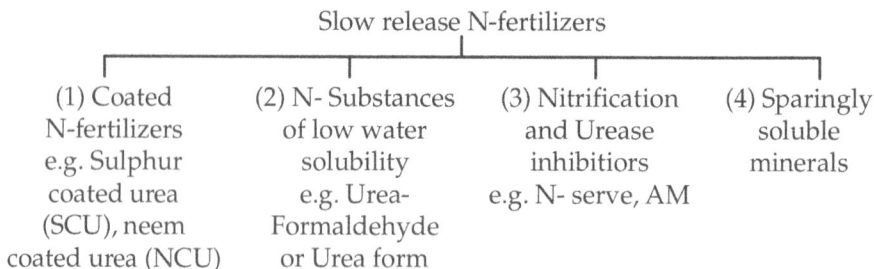

Slow release N-fertilizers

| (1) Coated N-fertilizers e.g. Sulphur coated urea (SCU), neem coated urea (NCU) | (2) N- Substances of low water solubility e.g. Urea-Formaldehyde or Urea form | (3) Nitrification and Urease inhibitiors e.g. N- serve, AM | (4) Sparingly soluble minerals |

29 *Classification of Phosphatic fertilizers*

(1) Water soluble monocalcium phosphate $[Ca(H_2PO_4)_2]$

e.g. Single Super Phosphate (SSP) = 6.8 – 7.74% P

Double Super Phosphate (DSP) = 13.76% P

Triple Super Phosphate (SSP) = 19.78 – 20.64% P

(2) Citric Acid Soluble, dicalcium phosphate $[Ca_3\ 2H_2(PO_4)_2\ OR\ CaHPO_4]$

e.g. Basic slag, silicates of lime = 6.02 – 7.74% P

Dicalcium phosphate = 14.62 – 16.77% P

(3) Phosphatic fertilizers not soluble in water or not soluble in citric acid tricalcium phosphate $[Ca_3\ (PO_4)_2]$

e.g., Rock phosphate = 8.6 – 17.2% P

Raw bonemeal = 8.6 – 10.75% P

Steamed bonemeal = 9.46% P

30. *Types of erosion*

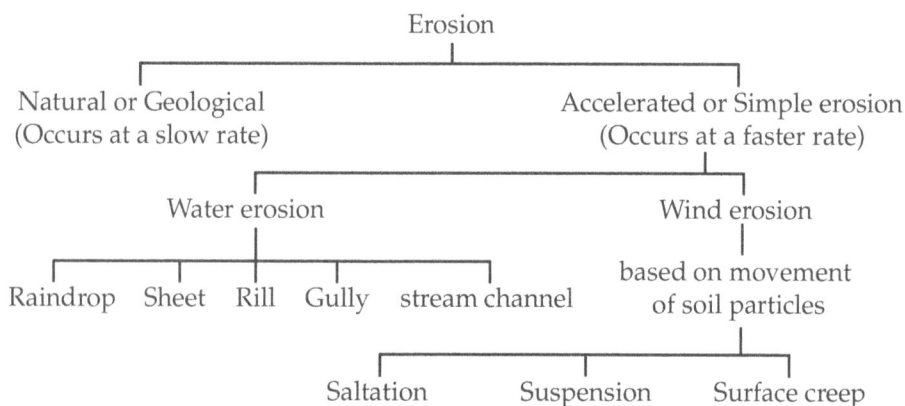

Erosion

| Natural or Geological (Occurs at a slow rate) | Accelerated or Simple erosion (Occurs at a faster rate) |

Water erosion — Wind erosion

Raindrop   Sheet   Rill   Gully   stream channel

based on movement of soil particles

Saltation   Suspension   Surface creep

6

DIAGRAMMATIC REPRESENTATIONS AND SHORT NOTES

1. Volume composition of Soil

Soil contains about 50 per cent solid space and 50 per cent pore space. The total solid space of the soil is occupied by mineral matter and organic matter by about 45 per cent and 5 per cent respectively. The total pore space of the soil is occupied by air and water on 50:50 bases, i.e., in this case 25 per cent air and 25 per cent water.

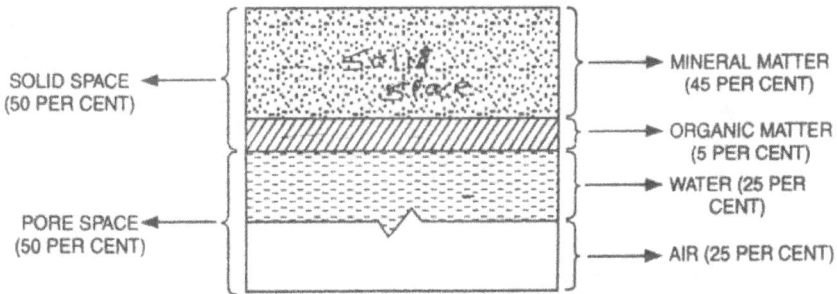

Figure 1 : Volume composition of soil suitable for plant growth. The amount of air and water will fluctuate depending on the weather and other factors (*Source: Dagi et al., 1996*).

2. Theoretical Soil profile consisting of all horizons

Organic Horizons : The 'O' group are the organic horizons

(1) **OL_1:** In this layer, the original forms of plants and animals can be easily identified by naked eye.

(2) **OL_2:** In this layer, the original forms of plants and animals cannot be recognized virtually

'A' (Eluvial) Horizons : It is characterized as zones of washing out or maximum leaching.

(1) **A_1 :** Uppermost mineral soil horizon rich in organic matter which impacts darker colour.

(2) A_2: The horizon of washing out of eluviations of clay, iron and aluminium oxides.

(3) A_3: It is transition horizon between A and B having properties more alike those of A_1 or A_2 that of B horizons. Sometimes it may be absent.

'B' (Illuvial) Horizons

These horizons are the zone of 'washing in' or accumulation of material such as iron and aluminum oxides and silicate clays.

(1) B_1: It is a transition horizon between A and B having properties more nearly like B than A. It may be sometimes absent.

(2) B_2: Zone of maximum accumulation of clays and hydrous oxides. Development of blocky or prismatic structure is found.

(3) B_3: It is the transition horizon between B and C having properties more like those of B than C horizon.

'C' Horizon

It is the unconsolidated material underlying the solum (A plus B horizon). This horizon also considered as outside the zones of major biological activities.

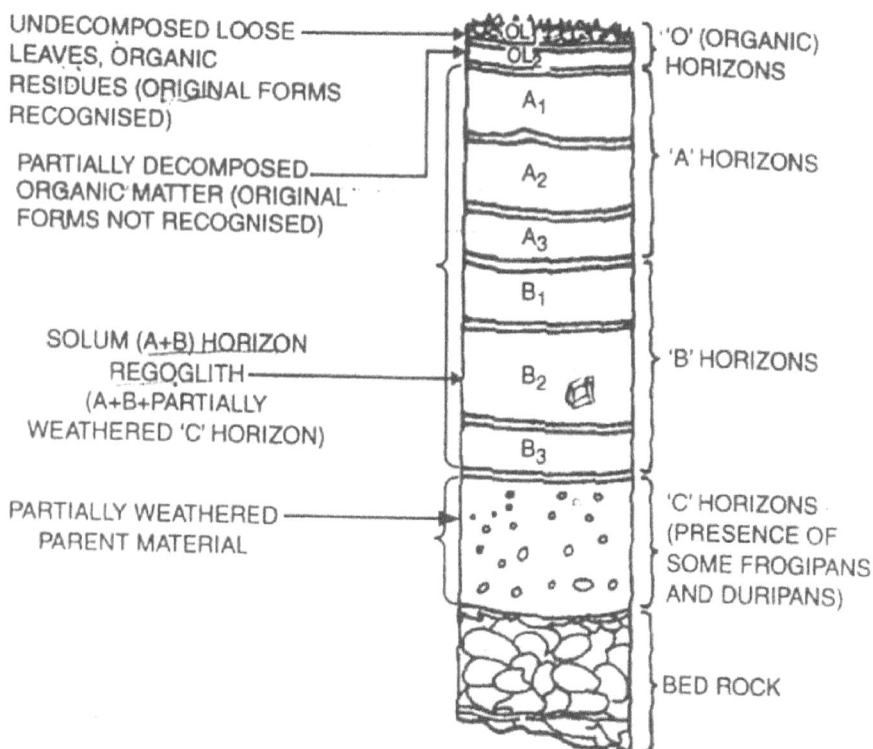

Figure 2 : Theoretical soil profile consisting of all horizons (*Source: Dagi et al., 1996*).

3. Weathering sequence by Goldich (1938)

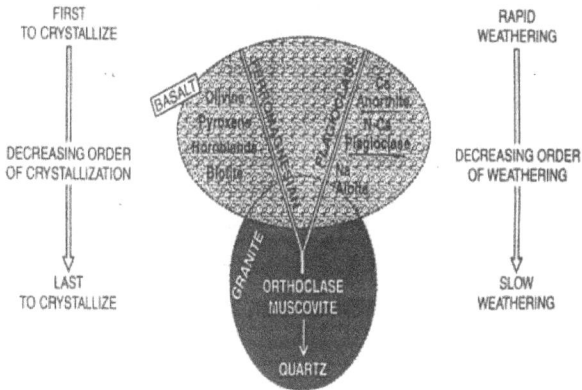

Figure 3 : Relative rate of chemical weathering of some common igneous rock-forming minerals (*Source: Dagi et al., 1996*).

4. Structure of water

Water is simple compound its individual molecules containing one oxygen atom and two much smaller hydrogen atoms. The elements are bonded together covalently, each hydrogen or proton sharing its single electron with the oxygen instead of the atoms are attached to the oxygen as a 'V' shaped arrangement and are separated from each other by angle of only 105°.

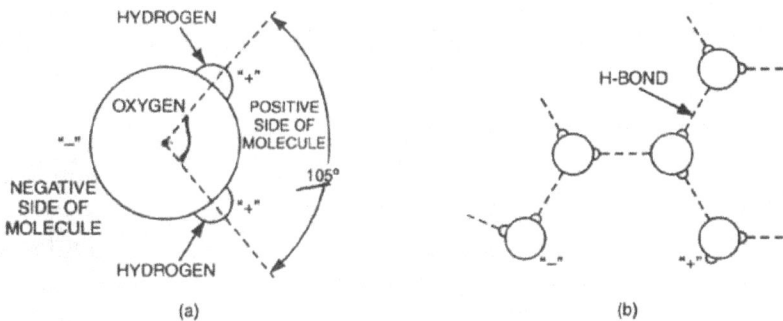

Figure 4 : (*a*) The polar water molecules, (*b*) the bonding of water to itself through H-bonding is shown (*Source : Dagi et al., 1996*).

5. Orientation of water molecules on the surface of clay micelle and cation

Strong combined adhesion and cohesion forces cause water films of considerable thickness to be held on the surface of soil particles. The mechanism of adsorption of water on the soil surfaces are related to the adhesion and cohesion forces through hydrogen bonding and also related to the hydration of exchangeable ions which may result in some of them dissociating from the surface into the water. The effect of the cation on the water molecules is greater. The greater its charge and the smaller its size, so the greater its surface charge density, and these effects are influenced by the relative moisture content of the clay, by the heat evolved during wetting of clays and by the greater apparent density of the days in water.

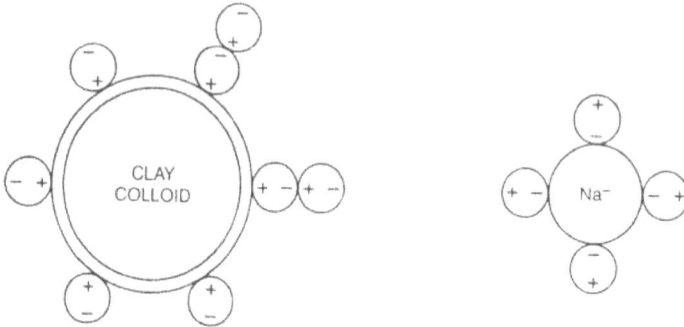

Figure 5 : Orientation of water molecules on the surface of clay micelle and cation (*Source: Dagi et al., 1996*).

6. Soil water constants and their approximate equivalents in bars of water potential as they affect the relative availability of water to plants

At water potential of 1/3 bares, water is held too loosely to overcome the effect of gravity and drains away. Capillary water (that held by capillary pressure) remains for plant use; this is held by water potentials ranging from –1/3 bar (field capacity) to –31 bars or lower (hygroscopic), depending upon the pore sizes of the soil. Plants can only use capillary water held by not more than –15 bars (the point of permanent wilting). Soil water at the air-dry state is held by water potentials that vary from –1,000 to –300 bars, depending on humidity.

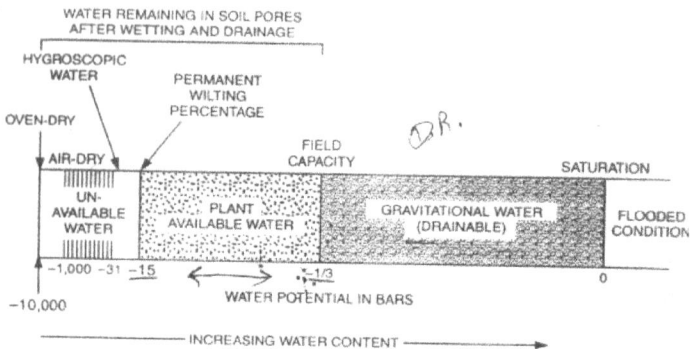

Figure 6 : Soil water constants and their approximate equivalents in bars of water potential as they affect the relative availability of water to plants. (*Source: Dagi et al., 1996*).

7. Electrical conductivity method for measurement of soil moisture

This method is based upon the changes in electrical conductivity with the variation in soil moisture. A gypsum block inside of which are two electrodes at a definite distance apart are used. These gypsum blocks require calibration for uniformity before use. The blocks are buried in the soil at the desired depth and the conductivity is measured with a modified Wheatstone bridge. Thought this method, the percentage of moisture from the field capacity to the wilting percentage can be easily measured. The limitation of this method is that it cannot be used in soils containing high salt concentration which interferes during the measurement. This method is practiced in the laboratory.

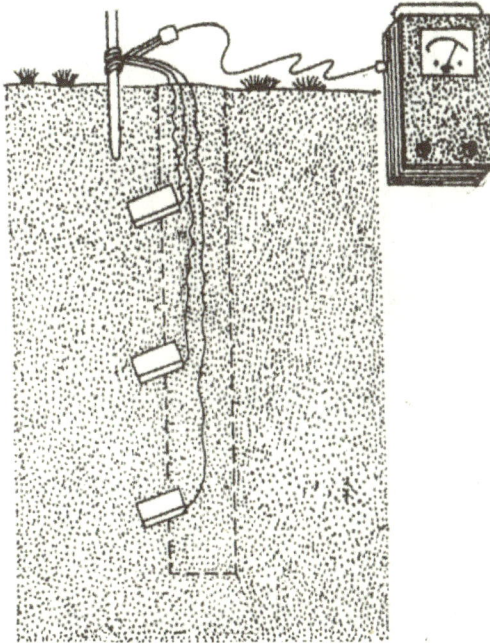

Figure 7 : Measuring electric conductivity with the Bouyoucos bridge
(*Source: Dagi et al., 1996*).

8. Measurement of soil moisture by using tensiometers

Tensiometers measure the metric potential of soil moisture in situ (filed) with the use of a porous clay cup attached to a tube filled with water. The water in the cup and tube is attached to a vacuum gauge or a mercury manometer. As the soil dries, water moves out through the porous cup, creating a suction or vacuum on the water column. These suction readings are then calibrated on the gauge to a specific soil to interpret the percent of moisture. Tensiometers can be used to schedule irrigation by placing one instrument at a depth of maximum root density and activity; a second instrument may be placed near the bottom of the active root zone. The main limitation of tensiometers is that they do not measure soil metric potential values as low as the usual wilting values. The actual range of effective measurement is only from 0 to -0.85 bars.

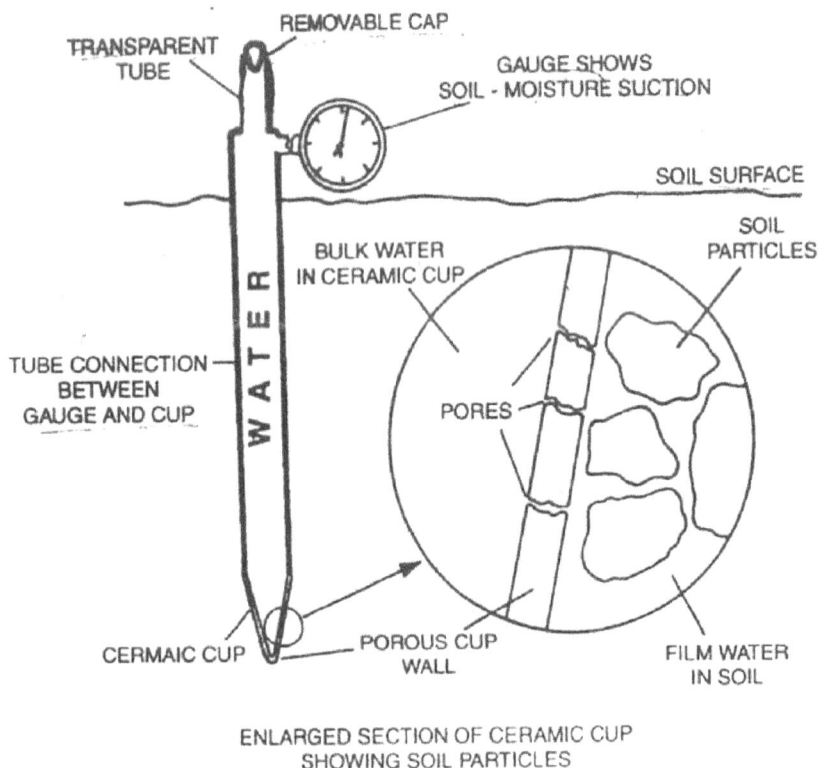

ENLARGED SECTION OF CERAMIC CUP
SHOWING SOIL PARTICLES

Figure 8 : Measuring soil moisture with tensiometer (*Source: Dagi et al., 1996*).

9. Type clay minerals

The most important mineral in this type, commonly found in soils, is kaolinite. The chemical composition of kaolinite is $Si_4Al_4O_{10}$ $(OH)_8$. The two sheets of each crystal unit of kaolinite are held together by oxygen atoms which are mutually shared by the silicon and aluminum atoms in their respective sheets. These units are held together very rigidly by hydrogen bonding (-H-) to the oxygen plane of the adjacent layer. So the lattice is fixed and kaolinite mineral does not allow water to penetrate between the layers and has almost no plasticity, cohesion, shrinkage and swelling properties. For example, Kaolinite, hallosite, anauxite and dickite.

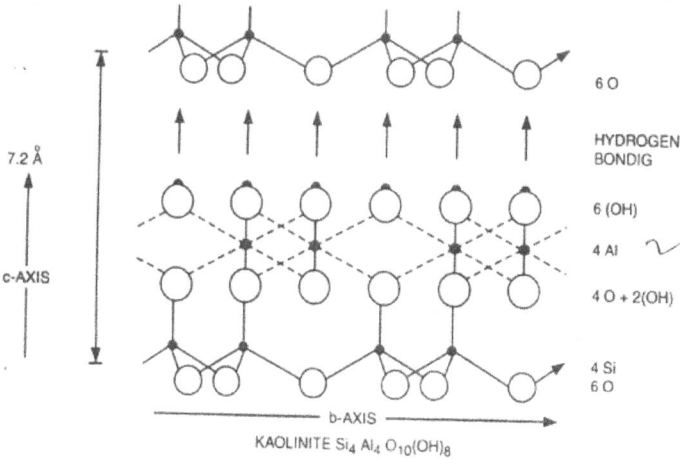

Figure 9 : Structure of Kaolinite (1:1 layer silicate) mineral (*Source: Dagi et al., 1996*).

10. Type clay minerals (expanding lattice type)

There are some important minerals in this type which includes montmorillonite (smectite group), vermiculite and other smectite group of minerals like beidellite, nontronite and saponite etc. The flake like crystals of this mineral are composed of 2:1 type crystal units. These crystal units are loosely held together by very weak oxygen to oxygen. Water molecules as well as cations are attracted between crystal units, causing expansion of the crystal lattice. The spacing (C-Axis) of the layers ranges from 12 to 18A (1.2 – 1.8 nanometers) and is variable with the exchangeable cation species and the degree of inter layer salvation. Montmorillonites are the swelling and sticky clays. Internal surface, cohesion and plasticity of this mineral are also very high.

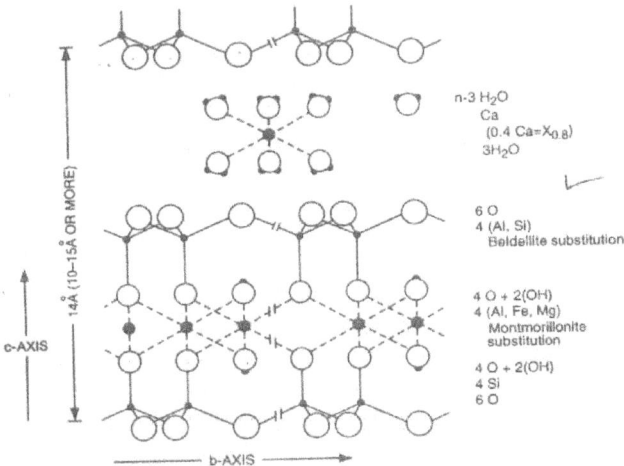

Figure 10 : Ions structures in the soil (*Source: Dagi et al., 1996*).

11. Non-expanding type clay minerals

In this group, hydrous mica or illite is the most important in soils. Like montmorillonite, illite has a 2:1 type lattice. About 15% of silion in silica sheets is substituted by aluminium. The excess of negative charge is satisfied largely by potassium in the interlattice layers, thus making the lattice structure non-expanding type. Thus hydrous mica has slight to moderate swelling.

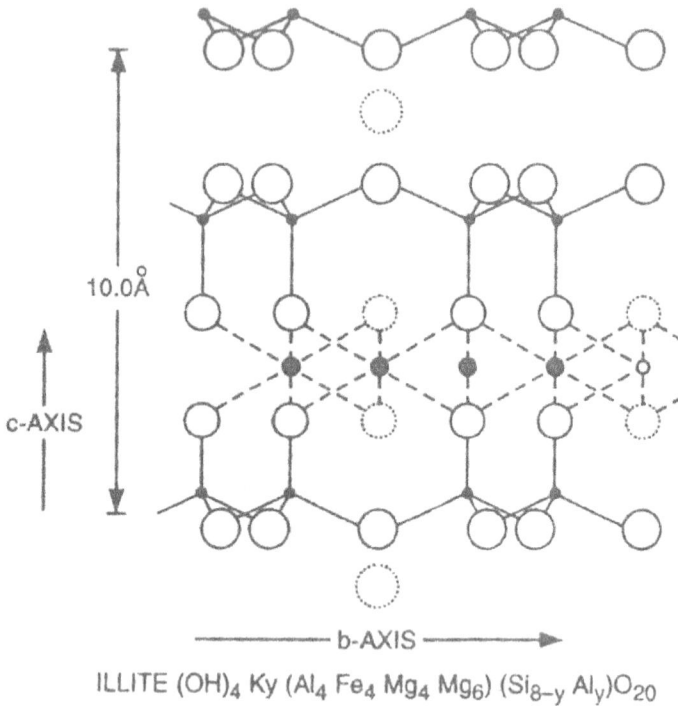

ILLITE $(OH)_4$ Ky $(Al_4 Fe_4 Mg_4 Mg_6)$ $(Si_{8-y} Al_y)O_{20}$

Figure 11 : Structure of illite (2:1 non-expenditure type) mineral
(*Source: Dagi et al., 1996*).

12. Type clay minerals

Chlorite (2:2 or 2:1:1 layer silicates) occurs extensively in soils chlorites are basically silicates of magnesium with some iron and aluminium present and it is composed of alternate talc and brucite layers. Chlorite mineral is similar to the unit lattice of vermiculite, except the hydrated mg in vermiculite is a firmly bonded magnesium hydroxide octahedral sheet. Thus, a layer of chlorite has 2 silica tetrahedral, aluminum octahedral and a magnesium octahedral sheet (2:2 or 2:1:1). Chlorite does not swell on wetting and bas low cation exchange capacities. It is almost non-expanding type of mineral because of its very little water adsorption.

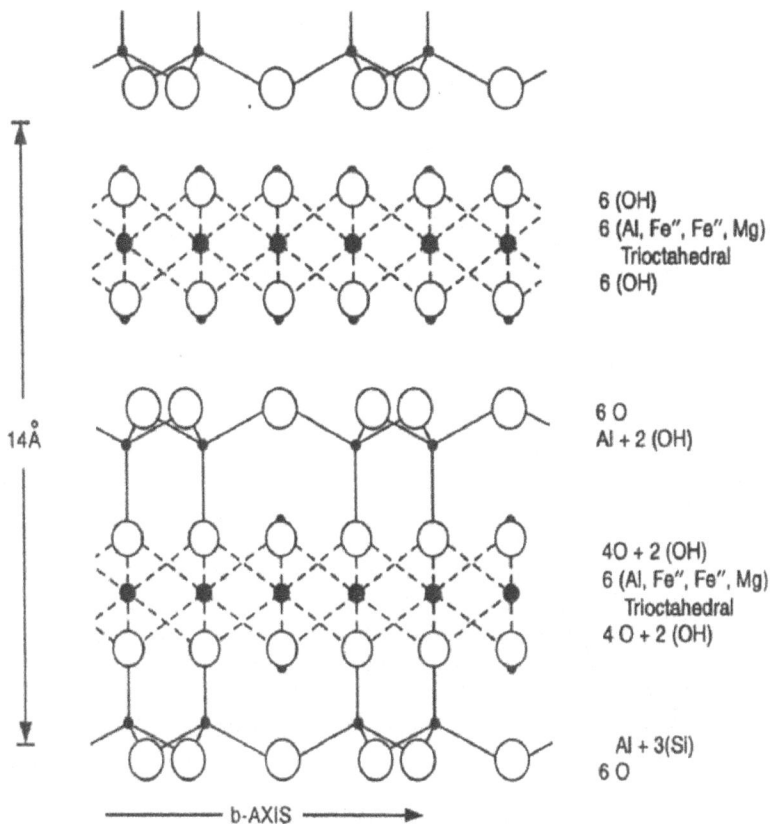

Figure 12 : Structure of chloride (2:2 layer silicate) mineral (*Source: Dagi et al., 1996*).

13. Changing cation exchange capacity

The cation exchange capacity of a soil changes with a change in pH (acidity or basicity). Most of the negatively charged exchange sites are from the isomorphous substitutions and are a permanent charge, sesquioxides (metal oxides) and kaolinite clay have only a few lattice sites with isomorphous substitution. The various hydroxyls of clays, humus and organic acid do ionize H^+ into the soil solution, thereby producing negatively charged cation exchange sites on these soil particles.

In acid solutions (high H^+ concentration); fewer H^+ ionize off the -0-. IN basic solutions fewer H^+ are in solution to adsorb to the -0- and more ionize of the -0-, leaving more cation exchange sites. In stronger acid solutions fewer hydroxyls (-OH) have the H^+ ionized, thereby leaving fewer pH-dependent CEC sites ionized. A relationship between soil pH and negative charge exchange sites is given in figure 13.

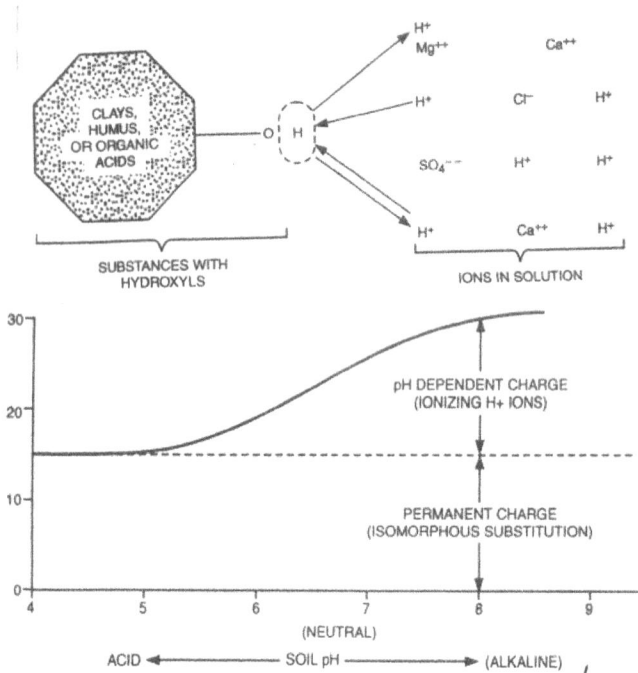

Figure 13(a) : Mechanisms for the development of negatively charged cation exchange site, **(b)** relationship between soil pH and negative charge exchange sites (*Source: Dagi et al., 1996*).

14. Elementary concepts of diffuse double layer

The exchangeable ions are surrounded by as forming a solution which is often a micellar solution or inner solution. The solution containing free electrolytes called outer solution or intermicellar solution. The ionic conditions on the outside surface of a clay particle or a pocket of clay particles dispersed in water or an electrolyte solution are controlled by the proportion of the exchangeable cations that disperse into the solution. A clay surface probably behaves as an effectively uncharged surface if the negative charge on the lattice is neutralized by monovalent cations that are tightly bound to the surface cations that are tightly bound to the surface but if the cations are hydrated a proportion but if the cations are hydrated a proportion tend to dissociate from the surface and will cause an electrical potential gradient to be set up near the surface. The system clay lattice-exchangeable cations-solution can be locked upon as forming a complex electrical double layer, known as the helm bolts double layer, the inner layer being the surface of the lattice carrying the negative charge and the outer layer being composed of two parts-a positive layer due to the cations bound to the lattice surface known as the fixed layer or stern layer, and a positive layer diffused in the solution close to the lattice surface known as the gouge diffuse layer.

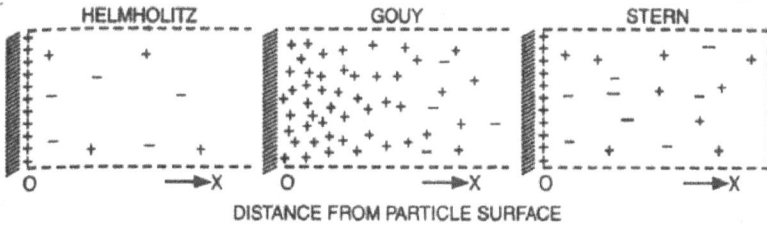

Figure 14 : Schematic representation of ion and potential distribution in the double layer according theories of Helmholtz, Gouy and Stern (*Source: Dagi et al., 1996*).

15. Ranges in soil reaction

For mineral soils the extreme range in soil reaction or pH extends from 3.5 to 10.5 sometimes in peat soils, soil reaction or pH may be low to 3.0 or less and some al kali soils the reaction or pH may be high as 11.0. The ranges in soil pH for most mineral soils are shown in figure.

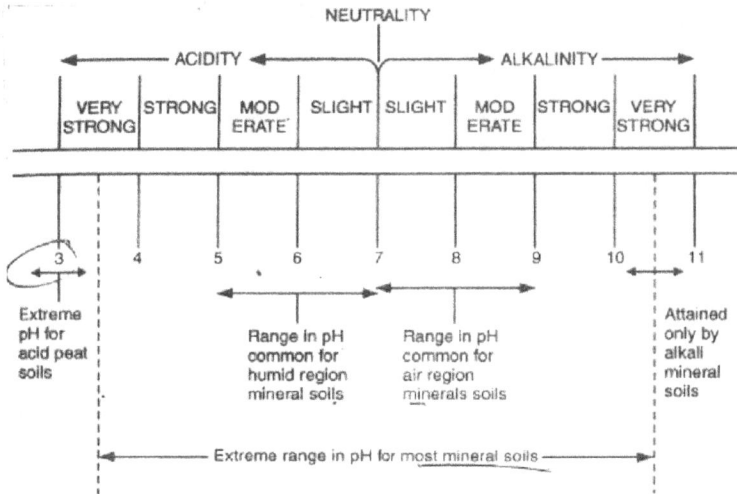

Figure 15 : Extreme range in pH for most mineral soils and the range commonly humid region (*Source: Dagi et al., 1996*).

16. Influence of soil reaction on nutrient availability and plant growth

Plants can grow on soils with a wide range of soil reaction of pH as well as diversified types of soil conditions. The diversity in plants growing on soils with a varying range of pH is the best evidence. Soil reaction and plant growth are interrelated as a result of the significant effect of soil reaction on soil environmental influence arising from increase or decrease in availability of nutrient element or from increasing concentration of certain plant nutrients to the toxic level, e.g., of Al^{3+}, Fe^{2+} or Mn^{2+} in acid soils. The other environmental factor is soil physical condition which is rendered unfavorable under low or high pH. Soil reaction (pH) is the most important factor which ovens availability of various essential as well as functional elements in soil by influencing the various soil properties like, physical, chemical and biological etc. A broad generalization of soil reaction and availability of nutrient elements in organic soils as well as mineral soils are shown in figures. From the figures it is found that the primary and secondary nutrients-nitrogen, phosphorus, calcium and magnesium are as available as or more available at a pH of 5.5 and 6.5 for organic and mineral soils than at any other pH. However, molybdenum, boron and copper availabilities are also relatively high at a range of pH 5.5 to 6.5. The micronutrients like iron, manganese and zinc are less available at pH of 5.5 and 6.5 than at more acidic reactions.

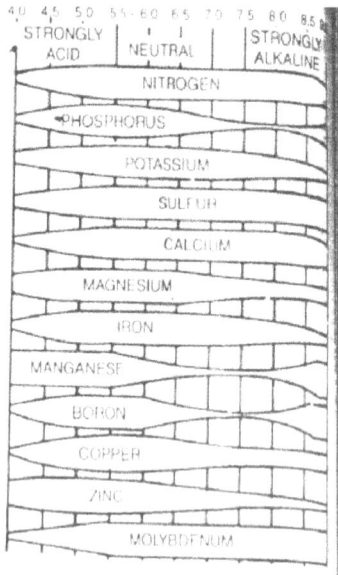
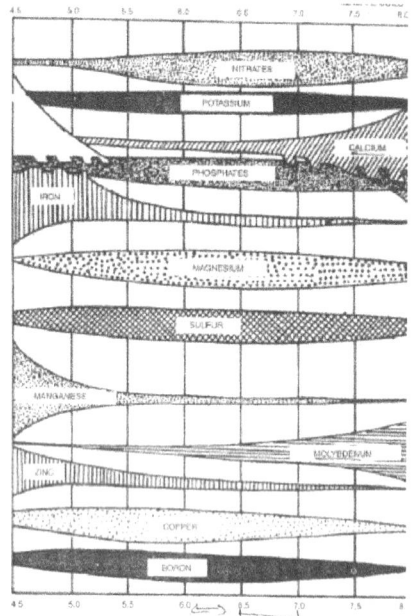

(a)   (b)

Figure 16 (a) : Availability of plant nutrients in relation to soil pH in organic soils, (b) Relationship between pH of the mineral soils and availability of nutrient elements (*Source: Dagi et al., 1996*).

17. Liming reaction to ameliorate the soil acidity

In this way, hydrogen ions (H⁺) in the soils solution react to form weakly dissociated water, and the calcium (Ca^{2+}) ion from limestone is left to undergo cation exchange reactions. The acidity of the soil is, therefore, neutralized and the per cent base saturation of the colloidal material is increased. The process of changing pH by the addition of lime [Ca (OH)₂] is illustrated in figure.

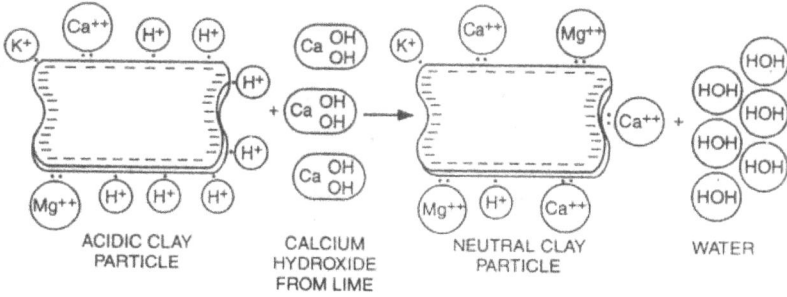

Figure 17 : Change of pH with the application of lime (*Source: Dagi et al., 1996*).

18. *Relationships among sodium adsorption ratio (SAR), exchangeable sodium percentage (ESP) and different salt affected soils*

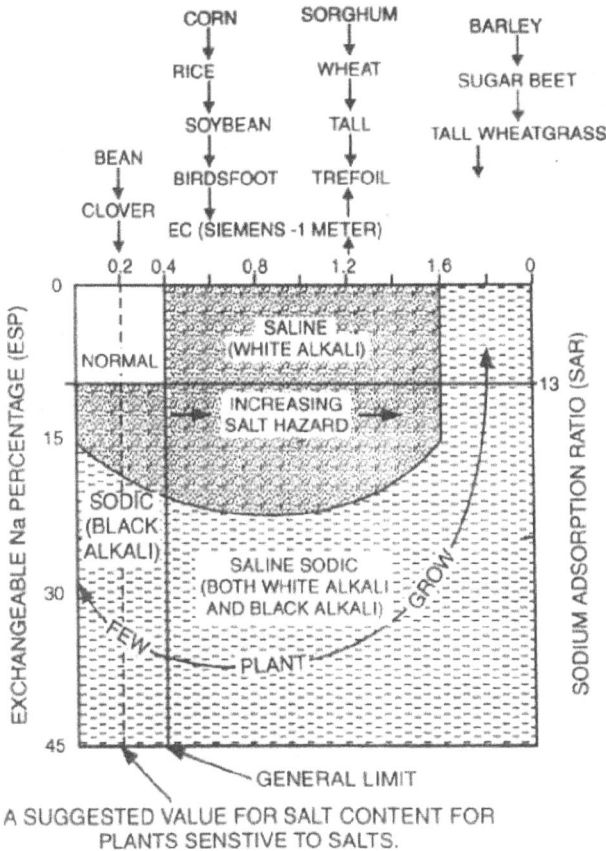

Figure 18 : Relationship among SAR, ESP and different salt affected soils
(*Source: Dagi et al., 1996*).

19. Leaching requirement in relation to salt tolerance of crops

In applying this equation, a value is usually assumed for ECdr (8 dSm^{-1} for most of the field crops) to represent the maximum soil salinity that can be tolerated. For irrigation waters with conductivities of 1, 2 and 3 Dsm^{-1} respectively, the leaching requirement will be 13, 25 and 38 per cent (taking the value of ECdw as 8 Dsm^{-1}). Leaching requirement as related to crop salt tolerance is shown in figure.

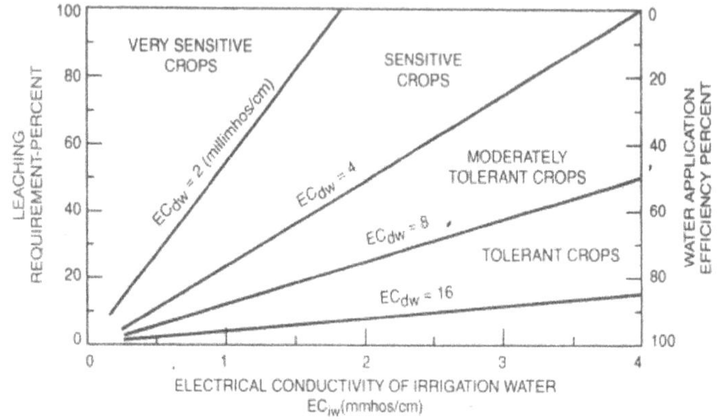

Figure 19 : Leaching requirement in relationship to salt tolerance of crops
(*Source: Dagi et al., 1996*).

20. Mechanism of release of fixed potassium

The activity of K$^+$ ions in soil solution around mica particles is a factor in determining the release of K$^+$ from micas. When the K$^+$ activity is less than the critical, K$^+$ is replaced from the inter layer by other cations from the solution. However, if the K level is greater than the critical value, the expanding 2:1 mica mineral takes K$^+$ from micas by removing the reaction products. Hence, leaching accelerates the transformation of micas to expansible 2:1 layer silicates and other weathering products. The mechanism of release of fixed K$^+$ by minerals is shown in figure.

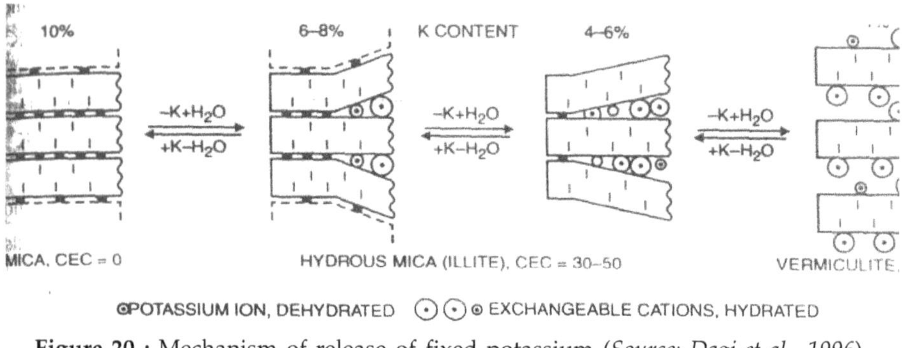

Figure 20 : Mechanism of release of fixed potassium (*Source: Dagi et al., 1996*).

21. Decomposition of organic compounds

When fresh organic materials (plants and animals) and in composited into the soil, three separate processes occur simultaneously as follows:

(a) Plant and animal tissue constituents disappear under the influence of microbial enzymes and appear new biological tissue, new microbial cells brining about an increase in soil of proteins, polysaccharides and nucleic acids etc.

(b) Breakdown of so formed organic compounds (according to their ease of decomposition) by different groups of micro-organisms and release of different essential plant nutrients like N, P, S etc. into the soil or immobilization takes place by a series of specific reactions.

(c) Formation of resultant compounds to microbial action from the original organic residues and also due to microbial synthesis.

A tentative scheme for the different stages of microbial decomposition of organic residues is shown in figure.

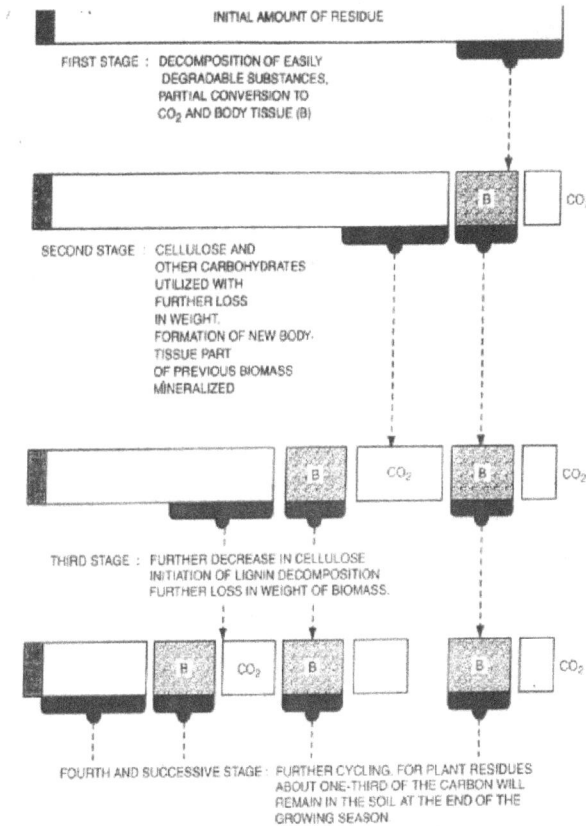

Figure 21 : Stages of decomposition of organic residues (*Source: Dagi et al., 1996*).

22. Carbon cycle

The carbon cycle revolves about CO_2 and its fixation and regeneration. Chlorophyll containing plants utilize the gas as their sole carbon source, and the carbonaceous matter thus synthesized serves to supply the animal world with performed organic carbon. Upon the death of the plant or animal, microbial metabolism assumes the dominant role in the cyclic sequence. The dead tissues undergo decay and are transformed into microbial cells and a large amount of heterogeneous carbonaceous compounds together known as humus or as the soil organic fractions. The cycle is completed and carbon made available with the final decomposition and production of carbon cycle is shown below.

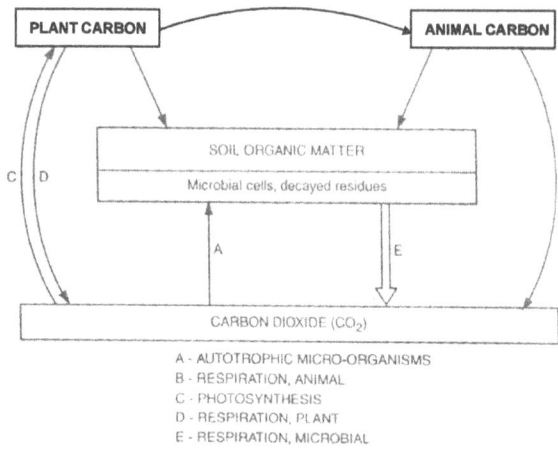

PLANT CARBON   ANIMAL CARBON

SOIL ORGANIC MATTER
Microbial cells, decayed residues

C   D

A   E

CARBON DIOXIDE (CO_2)

A - AUTOTROPHIC MICRO-ORGANISMS
B - RESPIRATION, ANIMAL
C - PHOTOSYNTHESIS
D - RESPIRATION, PLANT
E - RESPIRATION, MICROBIAL

Figure 22 : The carbon-cycle (*Source: Dagi et al., 1996*).

23. Concepts of nutrient availability

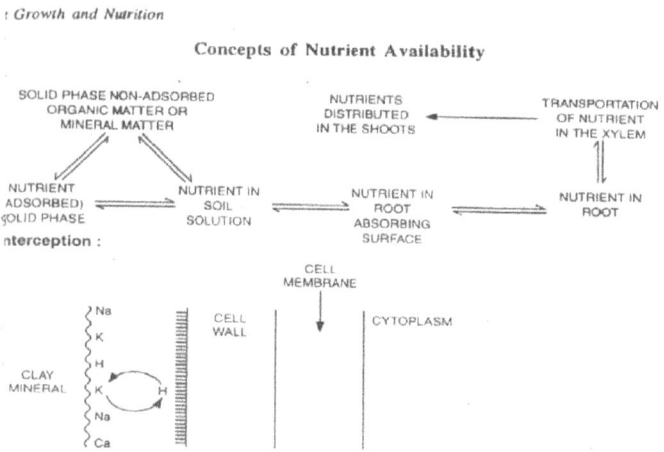

: *Growth and Nutrition*

Concepts of Nutrient Availability

SOLID PHASE NON-ADSORBED
ORGANIC MATTER OR
MINERAL MATTER

NUTRIENTS
DISTRIBUTED
IN THE SHOOTS

TRANSPORTATION
OF NUTRIENT
IN THE XYLEM

NUTRIENT
ADSORBED)
SOLID PHASE

NUTRIENT IN
SOIL
SOLUTION

NUTRIENT IN
ROOT
ABSORBING
SURFACE

NUTRIENT IN
ROOT

nterception :

CELL
MEMBRANE

Na
K
H
K   H
Na
Ca

CLAY
MINERAL

CELL
WALL

CYTOPLASM

Figure 23 : Nutrient availability concept (*Source: Dagi et al., 1996*).

24. *Mobility of soil metal ions boy chelates*

The chelated metals are protected against various soil sections, the nutrients remain in these combinations are considered available to associated with metallic cations is relatively higher association of cations. A simple mechanism showing the mobility of metal ions by chelates from soil to plant root surfaces is depicted diagrammatically.

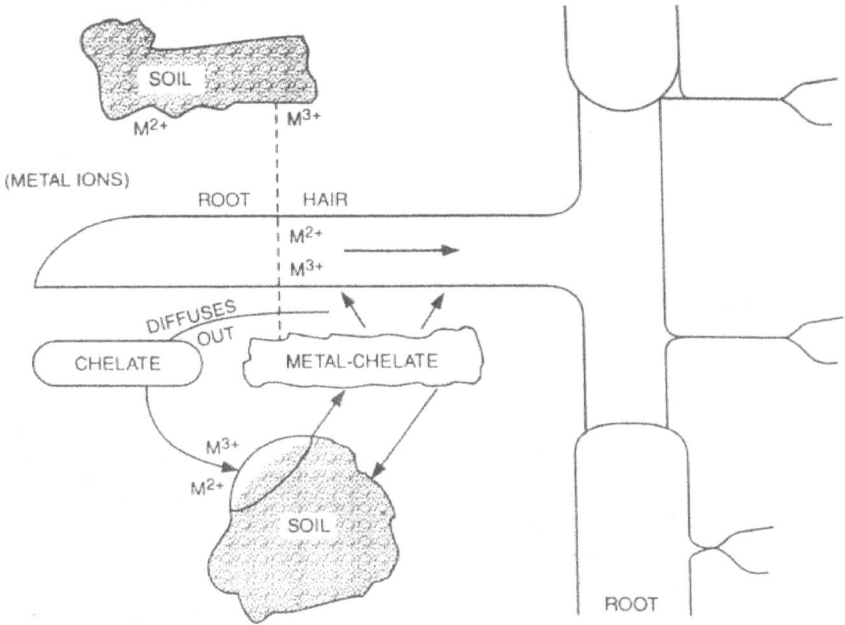

Figure 24 : Mobility of soil metal ions by chelates (*Source: Dagi et al., 1996*).

25. *Sulphur cycle*

Transformation of different paths of sulphur cycle is carried out by variety of micro-organisms as follows:

Path 2: Carried out by dissimilatory reduces, e.g., Desulphovirio, Desulphotomaculum.

Path 1 and 3: Assimilatory reducers like bacteria, fungi, algae and plant.

Path 5: Carried out by Desulphuromonas.

Path 4, 6 and 8: Carried out by chemolithotrophs (Thiobacillas Beggiator) and photo lithotrophs (Chlorobium and Chromatium)

Path 7 and 9: Carried out by heterotrophic micro-organisms, and chemo and photolithographs.

Biological sulphur cycle showing major chemical pools of sulphur proposed by Trudinger (1979) is depicted in figure.

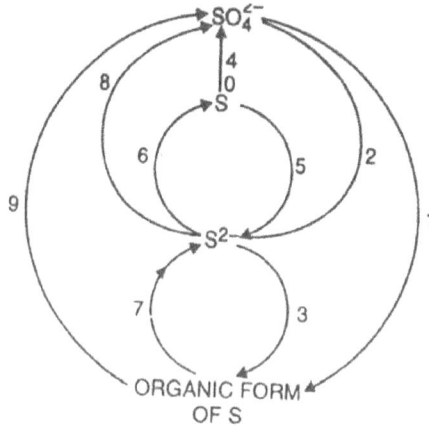

Figure 25 : Biological sulphur cycle (*Source: Dagi et al., 1996*).

26. Soil profile of a submerged soil

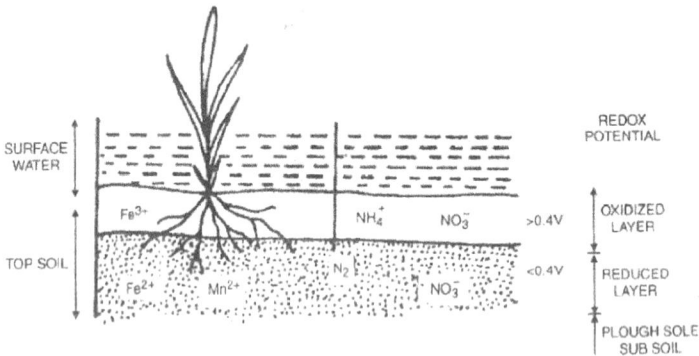

Figure 26 : Soil profile of a submerged soil (*Source: Dagi et al., 1996*).

27. Characteristics of oxidized and reduced zones in submerged soil

The low redox potential (Eh) resulting from soil submergence, influences the rice growth which are as follows:

(a) A low Eh harms germination and seedling emergence but not growth of the well established plant.

(b) Los Eh destroys NO^3-N but favours NH_4^+-N accumulation and nitrogen fixation and hence results net increase in the soils nitrogen regime.

(c) Low redox potential benefits rice by increasing the availability of N, P, Si, FE, Mn and Mo.

(d) It causes harmful effect to rice by decreasing the availability of S, Su and Zn.

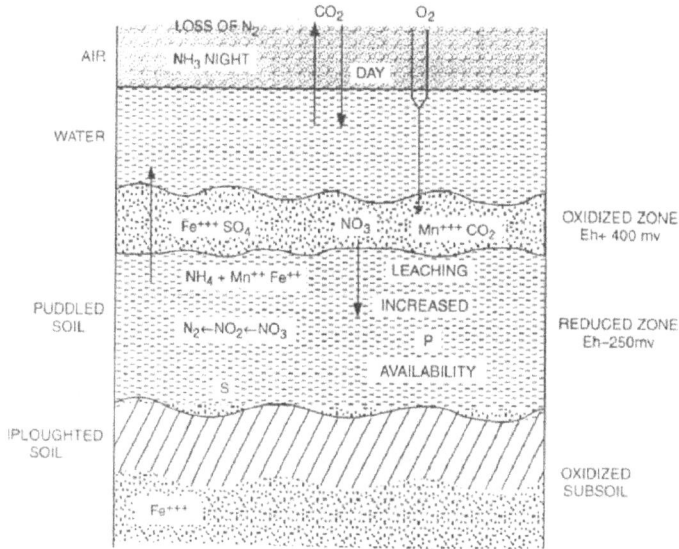

Figure 27 : Characteristics of oxidized and reduced zones in submerged soil
(*Source: Dagi et al., 1996*).

28. *Major Types of Soil Found in India?*

The committee appointed by the Indian Council of Agricultural Research (ICAR) classified the Indian soil in the following main groups:

(*i*)   Alluvial Soils   (*ii*)   Black Soils

(*iii*) Red Soils   (*iv*)   Laterite Soils

(*v*)   Mountain Soils   (*vi*)   Desert Soils

(*i*) Alluvial Soil

It is the most important type of soil found in India covering about 40 per cent of the total land area. It is very fertile and contributes the largest share of agricultural wealth. This soil supports nearly half of the Indian population. The alluvial soil is found mostly in the Northern Plains, starting from Punjab in the west to West Bengal and Assam in the east. It is also found in the deltas of the Mahanadi, Godavari, Krishna and Kaveri rivers in the Peninsular India. The northern parts and the coastal areas of Gujarat also have some deposits of alluvial soil. Many rivers originate from the Himalayan Mountains and bring a large amount of sediment with them. It is deposited in the river valleys and the flood plains. Thus, the parent material of the alluvial soils is always of transported origin.

The fine particles of sand, silt and clay are called alluvium. The alluvial soil can be divided into old alluvium, also called bangar, and new alluvium, called khadar. Remember, the new alluvium can be about ten thousand years old.

(*i*)   The new alluvium is deposited in the flood plains and deltas. These areas are flooded almost every year.

(*ii*) The old alluvium is found on the higher side of the river valleys, i.e. about 25 meters above the flood level.

(*iii*) The khadar soil is sandy and light in colour, while the bhangar soil is clayey and dark.

(*iv*) The khadar soil is more fertile than the bangar soil.

(*v*)   The alluvial soils contain adequate potash, phosphoric acid and lime.

(*vi*) They are generally deficient in organic and nitrogenous contents.

(*vii*) The old alluvium often contains lime nodules, known as kankar.

The fertility of the alluvial soil varies from place to place. Due to its softness and fertility, alluvial soil is most suited to irrigation and can produce bumper crops of rice, wheat, maize, sugar cane, tobacco, cotton, jute, oilseeds, etc.

(*ii*) Black Soil

The black soil is locally called regur, a word derived from Telugu word 'reguda'. It is also called the Black Cotton Soil, as cotton is the most important crop grown in this soil. The black soil is mostly found in the Deccan Trap, covering large areas of Maharashtra, Gujarat and western Madhya Pradesh. It is also found in some parts of Godavari and Krishna river valleys, covering parts of Karnataka, Andhra Pradesh and Tamil Nadu.

(*i*)   The black soil has been formed thousands of years ago, due to the solidification of volcanic lava.

(*ii*)   This soil is made up of extremely fine clayey material.

(*iii*) The black soil is well-known for its capacity to hold moisture.

(*iv*)   This soil is rich in calcium carbonate, magnesium carbonate, potash and lime, but poor in phosphoric content.

(*v*)   During the rainy season, black soil becomes sticky and is difficult to till as the plough gets stuck in the mud.

(*vi*)   During the hot dry season, the surface of this soil develops cracks.

(*vii*) These cracks help in the aeration of the soil.

(*viii*) Actually the black soil should be tilled immediately after the first or the pre-monsoon showers.

Generally, in the upland regions, the black soil has low fertility, while in the valleys or lowlands; this soil is darker, deeper and very fertile. Due to high fertility and capacity to hold moisture, black soil is widely used for producing cotton, wheat, linseed, millets, tobacco and oilseeds. With proper irrigation facilities, this soil can also produce rice and sugar cane.

(*iii*) Red Soil

The red soil occupies about 10 per cent area of India, mostly in the south-eastern part of the Peninsular India. This area encircles the entire black soil region. The red soil is found in Tamil Nadu, parts of Karnataka, southeast Maharashtra, eastern parts of Andhra Pradesh, Madhya Pradesh, Orissa and Jharkhand.

(*i*) Most of the red soil has been formed due to weathering of igneous and metamorphic rocks.

(*ii*) The red colour is due to the high percentage of iron contents.

(*iii*) The texture of the red soil varies from sandy to clayey, and the majority being loamy.

(*iv*) On the uplands, the red soil is thin, poor, and porous and has loose gravel.

(*v*) In the lower areas, the soil is deep, rich, fine grained and fertile.

(*vi*) This soil is rich in potash, but poor in lime, phosphate, nitrogen and humus.

With proper doses of fertilizers and irrigation the red soils can give excellent yields of cotton, wheat, rice, pulses, millets, tobacco, oilseeds, etc.

(*iv*) Laterite Soil

The word 'laterite' has been derived from a Latin word meaning 'brick'. The laterite soil is widely spread in India and is mainly found on the summits of the Western Ghats, Eastern Ghats, Rajmahal Hills, Vindhyas, Satpuras and Malwa plateau. It is well- developed in southern Maharashtra, and parts of Orissa, West Bengal, Karnataka, Andhra Pradesh, Kerala, Bihar, Assam and Meghalaya.

(*i*) The laterite soil is formed under conditions of high temperature and heavy rainfall with alternate wet and dry periods.

(*ii*) Such climatic conditions promote leaching of soil. Leaching is a process in which heavy rains wash away the fertile part of the soil.

(*iii*) The laterite soil is red in colour and composed of little clay and much gravel of red sandstones.

(*iv*) Laterite soil generally is poor in lime and deficient in nitrogen. The phosphate contents are generally high.

Due to intensive leaching, the laterite soil generally lacks fertility and is of low value for crop production. But when manure and timely irrigated, the soil is suitable for producing plantation crops like tea, coffee, rubber, coconut, arecanut, etc. It also provides valuable building materials.

(v) Mountain Soil

The mountain soil is generally found on the hill slopes covered with forests. In the Himalayan region such soil is mainly found in the valley basins, the depressions and the lesser steep slopes. The north-facing slopes generally support soil cover. Apart from the Himalayan region, this soil is also found in the Western and Eastern Ghats and in some parts of the Peninsular India.

(i)   The mountain soil is formed mainly due to the deposition of organic matter provided by the forests.

(ii) This soil is rich in humus, but poor in potash, phosphorus and lime.

(iii) It is heterogeneous in nature and varies from place to place.

(iv) The mountain soil is sandy with gravels and is porous.

For getting high yields of crops, heavy doses of fertilizers have to be applied. In the Himalayan region wheat, maize, barley and temperate fruits are grown on this soil. This soil is especially suitable for producing plantation crops, such as tea, coffee, spices and tropical fruits in Karnataka, Tamil Nadu and Kerala.

(vi) Desert Soil

The desert soil is found mostly in the arid and semi-arid regions, receiving less than 50 cm of annual rainfall. Such regions are mostly found in Rajasthan and the adjoining areas of Haryana and Punjab. The Rann of Kachchh in Gujarat is an extension of this region.

(i)   The sand in the desert areas is partly of local origin and partly being blown in from the Indus Valley.

(ii) It includes even the wind-blown loess.

(iii) The desert soil has sand (90 to 95 per cent) and clay (5 to 10 per cent).

(iv) In some regions this soil has high percentage of soluble salts, but lacks in organic matter.

(v)   The nitrogen content is low, but the phosphate content is as high as in normal alluvial soil.

When water is made available through irrigation, the desert soil can produce a variety of crops, such as wheat, millets, barley, maize, pulses, cotton, etc. Shortage of water in the arid regions is the main limiting factor for agriculture.

29. Soil acidity

Area

Out of 25 m. ha of problematic acid soils of India with below pH 5.5, north east region represents 54% of the total area whereas; it occupies only 8% of the total geographical area.

Causes of soil acidity

1. Natural process: high rainfall, acidic parent material, type of vegetation.

2. Human induced: Soil erosion accompanied by loss of top fertile acidic soils and exposure poor fertile subsoil. In few pockets acidity is also induced by application of nitrogenous fertilizers.

Problems

1. Toxicity of aluminium mainly.

2. Deficiency of bases (calcium , magnesium and potassium).

3. Low availability of phosphorus caused by the high fixation capacity of soil.

4. Reduction of soil biological activities

5. Impairment of nitrogen fixation by legumes and deficiency of molybdenum.

6. Iron and manganese toxicities in submerged soils.

Management

1. Liming of acid soils to raise the soil pH to 5.5 to neutralize toxic aluminum i.e. make the soil to suit the plant.

2. Growing of acid tolerant plants: select the plants to suit the acid soil condition.

3. Breeding of acid tolerant crops especially pulses, oilseeds and cereals.

Technology available

1. Ready Reckoner for acid soils to raise soil pH to 5.5 by liming.

2. Application of 500 kg lime in furrows to each crop.

3. Tolerant plants: pineapple, coffee, tea, rubber, sweetpotato, cassava, potato, rice, pigeonpea, fingermillet, buckwheat, ricebean, colocasia, allocasia, ginger, turmeric, etc.

Soil pH

The soil pH is a measure of the acidity or basicity in soils. pH is defined as the negative logarithm (base 10) of the activity of hydronium ions (H^+ or, more precisely, H_3O^+ aq) in a solution. In water, it nomally ranges from -1 to 14, with 7 being neutral. A pH below 7 is acidic and above 7 is basic. Soil pH is considered a master variable in soils as it controls many chemical processes that take place. It specifically affects plant nutrient availability by controlling the chemical forms of the nutrient. The optimum pH range for most plants is between 5.5 and 7.0, however many plants have adapted to thrive at pH values outside this range.

Denomination	pH range
Ultra acid	< 3.5
Extreme acid	3.5–4.4
Very strong acid	4.5–5.0
Strong acid	5.1–5.5
Moderate acid	5.6–6.0
Slight acid	6.1–6.5
Neutral	6.6–7.3
Slightly alkaline	7.4–7.8
Moderately alkaline	7.9–8.4
Strongly alkaline	8.5–9.0
Very strongly alkaline	> 9.0

Sources of Acidity

Acidity in soils comes from H^+ and Al^{3+} ions in the soil solution and sorbed to soil surfaces. While pH is the measure of H^+ in solution, Al^{3+} is important in acid soils because between pH 4 and 6, Al^{3+} reacts with water (H_2O) forming $AlOH^{2+}$, and $Al(OH)_2{}^+$, releasing extra H^+ ions. Every Al^{3+} ion can create 3 H^+ ions. Many other processes contribute to the formation of acid soils including rainfall, fertilizer use, plant root activity and the weathering of primary and secondary soil minerals. Acid soils can also be caused by pollutants such as acid rain and mine spoiling.

- **Rainfall:** Acid soils are most often found in areas of high rainfall. Excess rainfall leaches base cation from the soil, increasing the percentage of Al^{3+} and H^+ relative to other cations. Additionally, rainwater has a slightly acidic pH of 5.7 due to a reaction with CO_2 in the atmosphere that forms carbonic acid.

- **Fertilizer use:** Ammonium ($NH_4{}^+$) fertilizers react in the soil in a process called nitrification to form nitrate ($NO_3{}^"$), and in the process release H^+ ions.

- **Plant root activity:** Plants take up nutrients in the form of ions ($NO_3{}^-$, $NH_4{}^+$, Ca^{2+}, H_2PO_4, etc.), and often, they take up more cations than anions. However plants must maintain a neutral charge in their roots. In order to compensate for the extra positive charge, they will release H^+ ions from the root. Some plants will also exude organic acids into the soil to acidify the zone around their roots to help solubilize metal nutrients that are insoluble at neutral pH, such as iron (Fe).

- **Weathering of minerals:** Both primary and secondary minerals that compose soil contain Al. As these minerals weather, some components such as Mg, Ca, and K, are taken up by plants, others such as Si are leached from the soil, but due to chemical properties, Fe and Al remain in the soil profile. Highly weathered soils are often characterized by having high concentrations of Fe and Al oxides.

- **Acid Rain:** When atmospheric water reacts with sulfur and nitrogen compounds that result from industrial processes, the result can be the formation of sulfuric and nitric acid in rainwater. However the amount of acidity that is deposited in rainwater is much less, on average, than that created through agricultural activities.

- **Mine Spoil:** Severely acidic conditions can form in soils near mine spoils due to the oxidation of pyrite.

- Potential acid sulfate soils naturally formed in waterlogged coastal and estuarine environments can become highly acidic when drained or excavated.

- Decomposition of organic matter by micro organisms releases CO_2 which when mixed with soil water can form carbonic acid (H_2CO_3).

30. Soil basicity

Sources of basicity

Basic soils have a high saturation of base cations (K^+, Ca^{2+}, Mg^{2+} and Na^+). This is due to an accumulation of soluble salts which are classified as either saline soil, sodic soil, saline-sodic soil or alkaline. All saline and sodic soils have high salt concentrations, with saline soils being dominated by calcium and magnesium salts and sodic soils being dominated by sodium. Alkaline soils are characterized by the presence of carbonates. Soil in areas with limestone near the surface are alkaline from the calcium carbonate in limestone constantly mixing with the soil. Groundwater sources in these areas contain dissolved limestone.

Acid affected soils

Plants grown in acid soils can experience a variety of symptoms including aluminium (Al), hydrogen (H), and/or manganese (Mn) toxicity, as well as potential nutrient deficiencies of calcium (Ca) and magnesium (Mg). Aluminium toxicity is the most widespread problem in acid soils. Aluminium is present in all soils, but dissolved Al^{3+} is toxic to plants; Al^{3+} is most soluble at low pH, above pH 5.2 little aluminum is in soluble form in most soils. Aluminium is not a plant nutrient, and as such, is not actively taken up by the plants, but enters plant roots passively through osmosis. Aluminium damages roots in several ways: In root tips and Aluminium interferes with the uptake of Calcium, an essential nutrient, as well as bind with phosphate and interfere with production of ATP and DNA, both of which contain phosphate. Aluminium can also restrict cell wall expansion causing roots to become stunted. Below pH 4, H^+ ions themselves damage root cell membranes.

In soils with high content of manganese (Mn) containing minerals, Manganese toxicity can become a problem at pH 5.6 and below. Manganese, like aluminium becomes increasingly more soluble as pH drops, and Manganese toxicity symptoms can be seen at pH's below 5.6. Mn is an essential plant nutrient, so plants transport manganese into leaves. Classic symptoms of manganese toxicity are crinkling or cupping of leaves.

Nutrient availability in relation to soil pH

Nutrients needed in large amounts by plants are referred to as macronutrients and include nitrogen (N), phosphorus (P), potassium (K), calcium (Ca), magnesium (Mg) and sulfur (S). Elements that plants need in trace amounts are called trace nutrients or micronutrients. Trace nutrients are not major components of plant tissue but are essential for growth. They include iron (Fe), manganese (Mn), zinc (Zn), copper (Cu), cobalt (Co), molybdenum (Mo), and boron (B). Both macronutrient and micronutrient availability are affected by soil pH. In slightly to moderately alkaline soils, molybdenum and macronutrient (except for phosphorus) availability is increased, but P, Fe, Mn, Zn Cu, and Co levels are reduced and may adversely affect plant growth. In acidic soils, micronutrient availability (except for Mo and Bo) is increased. Nitrogen is supplied as ammonium (NH_4) or nitrate (NO_3) in fertilizer amendments, and dissolved N will have the highest concentrations in soil with pH 6–8. Concentrations of available N are less sensitive to pH than concentration of available P. In order for P to be available for plants, soil pH needs to be in the range 6.0 and 7.5. If pH is lower than 6, P starts forming insoluble compounds with iron (Fe) and aluminium (Al) and if pH is higher than 7.5 P starts forming insoluble compounds with calcium (Ca). Most nutrient deficiencies can be avoided between a pH ranges of 5.5 to 6.5, provided that soil minerals and organic matter contain the essential nutrients to begin with.

Increasing pH of acidic soil

The most common amendment to increase soil pH is lime ($CaCO_3$ or $MgCO_3$), usually in the form of finely ground agricultural lime. The amount of lime needed to change pH is determined by the mesh size of the lime (how finely it is ground)and the buffering capacity of the soil. A high mesh size (60–100) indicates a finely ground lime, that will react quickly with soil acidity. Buffering capacity of soils is a function of a soils cation exchange capacity, which is in turn determined by the clay content of the soil, the type of clay and the amount of organic matter present. Soils with high clay content, particularly shrink swell clay, will have a higher buffering capacity than soils with little clay. Soils with high organic matter will also have a higher buffering capacity than those with low organic matter. Soils with high buffering capacity require a greater amount of lime to be added than a soil with a lower buffering capacity for the same incremental change in pH. Other amendments that can be used to increase the pH of soil include wood ash, industrial CaO (burnt lime), and oyster shells. White firewood ash includes metal salts which are important for processes requiring ions such as Na^+ (sodium), K^+ (potassium), Ca^{2+} (calcium), which may or may not be good for the select flora, but decreases the acidic quality of soil. These products increase the pH of soils through the reaction of $CO_3^{2"}$ with H^+ to produce CO_2 and H_2O. Calcium silicate neutralizes active acidity in the soil by removing free hydrogen ions, thereby increasing pH. As its silicate anion captures H^+ ions (raising the pH), it forms monosilicic acid (H_4SiO_4), a neutral solute.

Decreasing pH of basic soil

- Iron sulphates or aluminium sulphate as well as elemental sulfur (S) reduce pH through the formation of sulfuric acid.

- Urea, urea phosphate, ammonium nitrate, ammonium phosphates, ammonium sulphate and monopotassium phosphate fertilizers.

- Organic matter in the form of plant litter, compost, and manure will decrease soil pH through the decomposition process. Certain acid organic matter such as pine needles, pine sawdust and acid peat are effective at reducing pH.

Saline soils

Alkali, or alkaline, soils are clay soils with high pH (> 8.5), a poor soil structure and a low infiltration capacity. Often they have a hard calcareous layer at 0.5 to 1 meter depth. Alkali soils owe their unfavorable physico-chemical properties mainly to the dominating presence of sodium carbonate which causes the soil to swell[1] and difficult to clarify/settle. They derive their name from the alkali group of elements to which the sodium belongs and that can induce basicity. Sometimes these soils are also referred to as (alkaline) sodic soils. Alkaline soils are basic, but not all basic soils are alkaline, see: "difference between alkali and base".

The causes of soil alkalinity are natural or they can be man-made

1. The natural cause is the presence of soil minerals producing sodium carbonate (Na_2CO_3) and sodium bicarbonate ($NaHCO_3$) upon weathering.

2. Coal fired boilers / power plants when using coal or lignite rich in limestone produces ash containing calcium oxide (CaO). CaO readily

dissolves in water to form slaked lime / $Ca(OH)_2$ and carried by rain water to rivers / irrigation water. Lime softening process precipitates Ca and Mg ions / removes hardness in the water and also converts sodium bicarbonates in river water into sodium carbonate.[2] Sodium carbonates (washing soda) further reacts with the remaining Ca and Mg in the water to remove / precipitate the total hardness. Also water soluble sodium salts present in the ash enhance the sodium content in water. The global coal consumption is 7700 million tons in the year 2011. Thus river water is made devoid of Ca and Mg ions and enhanced Na by coal fired boilers.

3. Many sodium salts are used in industrial and domestic applications such as Sodium carbonate, Sodium bicarbonate (baking soda), Sodium sulphate, Sodium hydroxide (caustic soda), Sodium hypochlorite (bleaching powder), etc. in huge quantities. These salts are mainly produced from Sodium chloride (common salt). All the sodium in these salts enter into the river / ground water during their production process or consumption enhancing water sodicity. The total global consumption of sodium chloride is 270 million tons in the year 2010. This is nearly equal to the salt load in the mighty Amazon River. Manmade sodium salts contribution is nearly 7% of total salt load of all the rivers.[3] Sodium salt load problem aggravates in the downstream of intensively cultivated river basins located in China, India, Egypt, Pakistan, west Asia, Australia, western USA, etc. due to accumulation of salts in the remaining water after meeting various transpiration and evaporation losses.

4. Another source of manmade sodium salts addition to the agriculture fields / land mass is in the vicinity of the wet cooling towers using sea water to dissipate waste heat generated in various industries located near the sea coast. Huge capacity cooling towers are installed in oil refineries, petrochemical complexes, fertilizer plants, chemical plants, nuclear & thermal power stations, centralized HVAC systems, etc. The drift / fine droplets emitted from the cooling towers contain nearly 6% sodium chloride which would deposit on the vicinity areas. This problem aggravates where the national pollution control norms are not imposed or not implemented to minimize the drift emissions to the best industrial norm for the sea water based wet cooling towers.

5. The man-made cause is the application of soft water in irrigation (surface or ground water) containing relatively high proportion of sodium bicarbonates and less calcium and magnesium.

Agricultural problems

Alkaline soils are difficult to take into agricultural production. Due to the low infiltration capacity, rain water stagnates on the soil easily and, in dry periods, cultivation is hardly possible without copious irrigated water and good drainage. Agriculture is limited to crops tolerant to surface water logging (e.g., rice, grasses) and the productivity is low.

Chemistry

Soil alkalinity is associated with the presence of sodium carbonate or washing soda (Na_2CO_3) in the soil, either as a result of natural weathering of the soil particles or brought in by irrigation and/or flood water. The sodium carbonate, when dissolved in water, dissociates into $2Na^+$ (two sodium cations, i.e. ions with a positive electric charge) and CO_3^{2-} (a carbonate anion, i.e. an ion with a double negative electric charge). The sodium carbonate can react with water to produce carbon dioxide (CO_2), escaping as a gas or absorbed by Algae, and sodium hydroxide (Na^+OH^-), which is alkaline (or rather basic) and gives high pH values (pH>8.5).

Solutions

Alkaline soils with solid $CaCO_3$ can be reclaimed with grass cultures, organic compost, waste hair / feathers, organic garbage, etc. ensuring the incorporation of much acidifying material (inorganic or organic material) into the soil, and enhancing dissolved Ca in the field water by releasing CO_2 gas. Deep plowing and incorporating the calcareous subsoil into the top soil also helps. Many times salts' migration to the top soil takes place from the underground water sources rather than surface sources. Where the underground water table is high and the land is subjected to high solar radiation, ground water oozes to the land surface due to capillary action and gets evaporated leaving the dissolved salts in the top layer of the soil. Where the underground water contains high salts, it leads to acute salinity problem. This problem can be reduced by applying mulch to the land. Using poly-houses during summer for cultivating vegetables/crops is also advised to mitigate soil salinity and conserve water / soil moisture. Poly-houses filter the intense summer solar radiation in tropical countries to save the plants from water stress and leaf burns.

Where the ground water quality is not alkaline / saline and ground water table is high, salts build up in the soil can be averted by using the land throughout the year for growing plantation trees / permanent crops with the help of lift irrigation. When the ground water is used at required leaching factor, the salts in the soil would not build up. Plowing the field soon after cutting the crop is also advised to prevent salt migration to the top soil and conserve the soil moisture during the intense summer months. This is done to break the capillary pores in the soil to prevent water reaching the surface of the soil.

Clay soils in high annual rain fall (more than 100 cm) areas do not generally suffer from high alkalinity as the rain water runoff is able to reduce/leach the soil salts to comfortable levels if proper rain water harvesting methods are followed. In some agricultural areas, the use of subsurface "tile lines" are used to facilitate drainage and leach salts. Continuous Drip irrigation would lead to alkali soils formation in the absence of leaching/drainage water from the field. It is also possible to reclaim alkaline soils by adding acidifying minerals like pyrite or cheaper alum or Aluminium sulfate.

Alternatively, gypsum (calcium sulfate, $CaSO_4$. $2H_2O$) can also be applied as a source of Ca^{++} ions to replace the sodium at the exchange complex. Gypsum also reacts with sodium carbonate to convert into sodium sulphate which is a neutral salt and does not contribute to high pH. There must be enough natural drainage to the underground, or else an artificial subsurface drainage system must be present, to permit leaching of the excess sodium by percolation of rain and/or irrigation water through the soil profile. Calcium Chloride is also used to reclaim alkali soils. $CaCl_2$ converts Na_2CO_3 into NaCl precipitating $CaCO_3$. NaCl is drained off by leaching water. Spent acids (HCl, H_2SO_4, etc.) can also be used to reduce the excess Na_2CO_3 in the soil. Where urea is made available cheaply to farmers, it is also used to reduce the soil alkalinity/salinity primarily. The NH_4 (Ammonium) present in urea which is a weak cation releases the strong cation Na from the soil structure into water. Thus alkali soils absorb/consume more urea compared to other soils.

To reclaim the soils completely one needs prohibitively high doses of amendments. Most efforts are therefore directed to improving the top layer only (say the first 10 cm of the soils), as the top layer is most sensitive to deterioration of the soil structure. The treatments, however, need to be repeated in a few (say 5) years time. Trees/plants follow gravitropism. It is difficult to survive in alkali soils for the trees with deeper rooting system which can be more than 60 meters deep in good non-alkali soils. It will be important to refrain from irrigation (ground water or surface water) with poor quality water. One way of reducing sodium carbonate is to cultivate glasswort or saltwort or barilla plants.

These plants sequester the sodium carbonate they absorb from alkali soil into their tissues. The ash of these plants contains good quantity of sodium carbonate which can be commercially extracted and used in place of sodium carbonate derived from common salt which is highly energy intensive process. Thus alkali lands deterioration can be checked by cultivating barilla plants which can serve as food source, biomass fuel and raw material for soda ash and potash, etc.

Leaching saline sodic soils

Saline soils are mostly also sodic (the predominant salt is sodium chloride), but they do not have a very high pH nor a poor infiltration rate. Upon leaching they are usually not converted into a (sodic) alkali soil as the Na^+ ions are easily removed. Therefore, saline (sodic) soils mostly do not need gypsum applications for their reclamation.

Drainage is the primary method of controlling soil salinity. The system should permit a small fraction of the irrigation water (about 10 to 20 percent, the drainage or leaching fraction) to be drained and discharged out of the irrigation project. In irrigated areas where salinity is stable, the salt concentration of the drainage water is normally 5 to 10 times higher than that of the irrigation water. Salt export matches salt import and salt will not accumulate. When reclaiming already salinized soils, the salt concentration of the drainage water will initially be much higher than that of the irrigation water (for example 50 times higher). Salt export will greatly exceed salt import, so that with the same drainage fraction a rapid desalinization occurs. After one or two years, the soil salinity is decreased so much, that the salinity of the drainage water has come down to a normal value and a new, favorable, equilibrium is reached. In regions with pronounced dry and wet seasons, the drainage system may be operated in the wet season only, and closed during the dry season. This practice of checked or controlled drainage saves irrigation water. The discharge of salty drainage water may pose environmental problems to downstream areas. The environmental hazards must be considered very carefully and, if necessary mitigating measures must be taken. If possible, the drainage must be limited to wet seasons only, when the salty effluent inflicts the least harm.

31. Different fertilizers used for agricultural crops

Fertilizer (or fertilizer) is any organic or inorganic material of natural or synthetic origin (other than liming materials) that is added to soil to supply one or more plant nutrients essential to the growth of plants. Conservative estimates report 30 to 50% of crop yields are attributed to natural or synthetic commercial fertilizer. Global market value is likely to rise to more than US$185 billion until 2019. The European fertilizer market will grow to earn revenues of approx. € 15.3 billion in 2018.

Mined inorganic fertilizers have been used for many centuries, whereas chemically synthesized inorganic fertilizers were only widely developed during the industrial revolution. Increased understanding and use of fertilizers were important parts of the pre-industrial British Agricultural Revolution and the industrial Green Revolution of the 20th century. Inorganic fertilizer use has also significantly supported global population growth — it has been estimated that almost half the people on the Earth are currently fed as a result of synthetic nitrogen fertilizer use.

Fertilizers typically provide, in varying proportions:

Six macro-nutrients: Nitrogen (N), phosphorus (P), potassium (K), calcium (Ca), magnesium (Mg), and sulfur (S);

Eight micro-nutrients: Boron (B), chlorine (Cl), copper (Cu), iron (Fe), manganese (Mn), molybdenum (Mo), zinc (Zn) and nickel (Ni).

The macro-nutrients are consumed in larger quantities and are present in plant tissue in quantities from 0.15% to 6.0% on a dry matter (0% moisture) basis (DM). Micronutrients are consumed in smaller quantities and are present in plant tissue on the order of parts per million (ppm), ranging from 0.15 to 400 ppm DM, or less than 0.04% DM. Only three other structural elements are required by all plants: carbon, hydrogen, and oxygen. These nutrients are supplied by water (through rainfall or irrigation) and carbon dioxide in the atmosphere.

32. What are the organic commercial fertilizers available in the market

Fertilizers are broadly divided into organic fertilizers (composed of organic plant or animal matter), or inorganic or commercial fertilizers. Plants can only absorb their required nutrients if they are present in easily dissolved chemical compounds. Both organic and inorganic fertilizers provide the same needed chemical compounds. Organic fertilizers provided other macro and micro plant nutrients and are released as the organic matter decays this may take months or years. Organic fertilizers nearly always have much lower concentrations of plant nutrients and have the usual problems of economical collection, treatment, transportation and distribution.

Inorganic fertilizers nearly always are readily dissolved and unless neither added have few other macro and micro plant nutrients nor added any 'bulk' to the soil. Nearly all nitrogen that plants use is in the form of NH_3 or NO_3 compounds. The usable phosphorus compounds are usually in the form of phosphoric acid (H_3PO_4) and the potassium (K) is typically in the form of potassium chloride (KCl). In organic fertilizers nitrogen, phosphorus and potassium compounds are released from the complex organic compounds as the animal or plant matter decays. In commercial fertilizers the same required compounds are available in easily dissolved compounds that require no decay they can be used almost immediately after water is applied. Inorganic fertilizers are usually much more concentrated with up to 64% (18-46-0) of their weight being a given plant nutrient, compared to organic fertilizers that only provide 0.4% or less of their weight as a given plant nutrient.

Nitrogen fertilizers are often made using the Haber-Bosch process (invented about 1915) which uses natural gas (CH_4^+) for the hydrogen and nitrogen gas (N_2) from the air at an elevated temperature and pressure in the presence of a catalyst to form ammonia (NH_3) as the end product. This ammonia is used as a feedstock for other nitrogen fertilizers, such as anhydrous ammonium nitrate (NH_4NO_3) and urea ($CO(NH_2)_2$). These concentrated products may be diluted with water to form a concentrated liquid fertilizer (e.g. UAN). Deposits of sodium nitrate ($NaNO_3$) (Chilean saltpeter) are also found the Atacama desert in Chile and was one of the original (1830) nitrogen rich inorganic fertilizers used. It is still mined for fertilizer.

In the Nitro phosphate or Odda Process (invented in 1927), phosphate rock with up to a 20% phosphorus (P) content is dissolved with nitric acid (HNO_3)

to produce a mixture of phosphoric acid (H_3PO_4) and calcium nitrate ($Ca(NO_3)_2$). This can be combined with a potassium fertilizer to produce a compound fertilizer with all three N:P:K: plant nutrients in easily dissolved form.

Phosphate rock can also be processed into water-soluble phosphate (P_2O_5) with the addition of sulfuric acid (H_2SO_4) to make the phosphoric acid in phosphate fertilizers. Phosphate can also be reduced in an electric furnace to make high purity phosphorus; however, this is more expensive than the acid process.

Potash can be used to make potassium (K) fertilizers. All commercial potash deposits come originally from marine deposits and are often buried deep in the earth. Potash ores are typically rich in potassium chloride (KCl) and sodium chloride (NaCl) and are obtained by conventional shaft mining with the extracted ore ground into a powder. For deep potash deposits hot water is injected into the potash which is dissolved and then pumped to the surface where it is concentrated by solar induced evaporation. Amine reagents are then added to either the mined or evaporated solutions. The amine coats the KCl but not NaCl. Air bubbles cling to the amine + KCl and float it to the surface while the NaCl and clay sink to the bottom. The surface is skimmed for the amine + KCl which is then dried and packaged for use as a K rich fertilizer KCl dissolves readily in water and is available quickly for plant nutrition.

Compound fertilizers often combine N, P and K fertilizers into easily dissolved pellets. The N:P:K ratios quoted on fertilizers give the weight percent of the fertilizer in nitrogen (N), phosphate (P_2O_5) and potash (K_2O equivalent)

The use of commercial inorganic fertilizers has increased steadily in the last 50 years, rising almost 20-fold to the current rate of 100 million tonnes of nitrogen per year. Without commercial fertilizers it is estimated that about one-third of the food produced now could not be produced. The use of phosphate fertilizers has also increased from 9 million tonnes per year in 1960 to 40 million tonnes per year in 2000. A maize crop yielding 6 - 9 tonnes of grain per hectare requires 31 - 50 kg of phosphate fertilizer to be applied, soybean requires 20 - 25 kg per hectare. Yara International is the world's largest producer of nitrogen based fertilizers.

Controlled-release types

Urea and formaldehyde, reacted together to produce sparingly soluble polymers of various molecular weights, is one of the oldest controlled-nitrogen-release technologies, having been first produced in 1936 and commercialized in 1955. The early product had 60 percent of the total nitrogen cold-water-insoluble, and the unreacted (quick release) less than 15%. Methylene ureas were commercialized in the 1960s and 1970s, having 25 and 60% of the nitrogen cold-water-insoluble, and unreacted urea nitrogen in the range of 15 to 30%. Isobutylidene diurea, unlike the methylurea polymers, is a single crystalline solid of relatively uniform properties, with about 90% of the nitrogen water-insoluble.

In the 1960s, the National Fertilizer Development Center began developing Sulfur-coated urea; sulfur was used as the principle coating material because of its low cost and its value as a secondary nutrient.[21] Usually there is another wax or polymer which seals the sulfur; the slow release properties depend on the degradation of the secondary sealant by soil microbes as well as mechanical imperfections (cracks, etc.) in the sulfur. They typically provide 6 to 16 weeks of delayed release in turf applications. When a hard polymer is used as the secondary coating, the properties are a cross between diffusion-controlled particles and traditional sulfur-coated.

Other coated products use thermoplastics (and sometimes ethylene-vinyl acetate and surfactants, etc.) to produce diffusion-controlled release of urea or soluble inorganic fertilizers. "Reactive Layer Coating" can produce thinner, hence cheaper, membrane coatings by applying reactive monomers simultaneously to the soluble particles. "Multicote" is a process applying layers of low-cost fatty acid salts with a paraffin topcoat. Besides being more efficient in the utilization of the applied nutrients, slow-release technologies also reduce the impact on the environment and the contamination of the subsurface water.

Application

Synthetic fertilizers are commonly used for growing all crops, with application rates depending on the soil fertility, usually as measured by a soil test and according to the particular crop. Legumes, for example, fix nitrogen from the atmosphere and generally do not require nitrogen fertilizer.

Studies have shown that application of nitrogen fertilizer on off-season cover crops can increase the biomass (and subsequent green manure value) of these crops, while having a beneficial effect on soil nitrogen levels for the main crop planted during the summer season. Nutrients in soil can be thrown out of balance with high concentrations of fertilizers. The interconnectedness and complexity of this soil 'food web' means any appraisal of soil function must necessarily take into account interactions with the living communities that exist within the soil. Stability of the system is reduced by the use of nitrogen-containing fertilizers, which cause soil acidification.

Applying excessive amounts of fertilizer has negative environmental effects, and wastes the growers' time and money. To avoid over-application, the nutrient status of crops should be assessed. Nutrient deficiency can be detected by visually assessing the physical symptoms of the crop. Nitrogen deficiency, for example has a distinctive presentation in some species. However, quantitative tests are more reliable for detecting nutrient deficiency before it has significantly affected the crop. Both soil tests and Plant Tissue Tests are used in agriculture to fine-tune nutrient management to the crops needs.

33. Background of organic farming worldwide

The organic agriculture Movement has progressed from small band of farmers having ideological orientation towards organic modes of production to national and international force with strong political influence. This paper makes an attempt to take a bird's eye view of the historical events of the organic agriculture movement and its present position.

In fact an organic agriculture is not alien to humanity but it is a reappearance of a long tradition in the context of contemporary food demands with due regards to quality of resource base in view of long term sustenance and ecological stability. The roots of organic agriculture movement are ancient but the current reformation is traceable to events initiated during the first half of the 20th century throughout the world. The organic agriculture movement has reappeared in the guise of many synonyms viz. organic farming, natural farming, sustainable agriculture, low-input sustainable agriculture (LISA), Biodynamic agriculture or just simply as alternative agriculture as perceived by many workers. Although there might be a little difference in the perception of meaning of these expressions one basic principle involved in all these expressions is to maximize agricultural production on sustainable basis without deterioration of the quality of basic resources and with due regards to ecological principles. Organic agriculture is holistic endeavor implying interactions between components such as crops with crops, crops with animals, soil conditions and fertility with pest and disease incidence in crop and livestock. Hence the key feature of organic agriculture is the interrelatedness of their components within the agro ecosystem. The farming practices consistent with the philosophy of organic agriculture may vary in their details but they have an important feature in common i.e. they are designed to drastically reduce (preferably to eliminate) the chemical pesticides and inorganic fertilizers which arc the key elements of modern agricultural system. This is achieved by use of organic manures, inclusions of legumes in rotation, recycling of crop residues and other organic wastes for supply of plant nutrients and also for improving soil health. For control of disease and pest adoption of integrated pest management is followed in which the use of pesticides that to in a minimum possible extent is considered as a last resort.

34. Historical background of organic farming

Plant growth has been the subject of curiosity to human being since the beginning of agriculture. The "substance" of plant has been debated since the time of early Greek philosophers and the current dichotomy between organic vs. inorganic sources of plant nutrients is not new. The earliest record of the benefits of green manure, animal manures dates back to Chou dynasty (-1100 BC) in China (Pieters, 1927). King (1911) reported that the practice of deliberately adding the organic matter to the soil dates back at least 4000 years and summarized his observations as the addition of organic matter is remarkable practice and it is only recently understood and added to the science of agriculture, namely the power of organic matter decaying in contact with, to liberate from it plant food.

Early Roman compilations of agricultural practices that enumerated the use of organic manures and crop rotations were accumulated by the agricultural observer Cais Plinius Secundus (23-79 AD), better known as pliny the Elder (Browne, 1943). However, it was Lord Walter Northbourne who was the first person to use the term "Organic Farming" in 1940 as a chapter heading in his book titled Look to the Land (Northbourne, 1940). But perhaps it is the Sir Albert Howard (1873-1947) who is considered as one of the leading advocates of organic farming of this century (Conford, 1988). Although he was born and educated in England, his most important work occurred in India where he was the Imperial Economic Botanist to the Indian Government from 1905 to 1924.and Director of the Institute of Plant Industry, Indore from 1924 to 1931. He summarized most of his research experiences and observations in a book "An Agricultural Testament" ultimately which brought him a reputation as on organic agricultural extremist. He attributed that plant and animal disease were due to unhealthy soils and if the soil was made healthy by organic technique, there would be no disease (Howard, 1943). This extreme view is still embraced by some proponents of organic agriculture. Despite the lack of scientific data to support such cause and effect relationship between soil health and disease, Howard championed this theory using testimonial and circumstantial evidences. For example, the extra ordinary health of the Hunza people in Himalaya was attributed by him to their consumption of organically grown food (Rodale, 1948). One of the Howard's most notable attacks in a book "An Agricultural Testament" was on the NPK mentality in agricultural research and held a view that only fertilizers are not the cure for all agricultural problems rather we should also consider soil biological and physical properties. He believed that loss of soil fertility seriously threatened the future of agriculture and that recycling of organic material through composting is a solution to avoid an agricultural catastrophe.

Between 1924 to 1931 Howard developed a process of compost making at Indore which and which is known Indore Process (Howard and Wad, 1931). He strongly advocated composting of organic waste as a source of plant nutrients and for improving soil properties. Although he was branded as an extremist in his view on organic farming during his lifetime more than 40 years after his death large segment of the society is moving closer to his view on agricultural sustainability via organic recycling.

35. What is status of organic movement in India?

Indian agriculture inherits the tradition of *Rishis kheti* (agriculture of sages) in ancient days where farming was practiced in a quest for realizing the underlaying unity of soil, plant and animals including human beings. The Rishis looked upon soil as a mother and ploughing as such was forbidden in order to protect it from damage. Fruits, tubers and milk were considered as the most appropriate diet for human being. Rice Millets and barley was grown in small quantities and used as offering to the sacrificial fire and leftover was used as food called *Prasad* and eaten as such. This was the type of farming perfectly in harmony with nature. Since then the Indian agriculture has travelled a long way to present day agriculture once again looking back to the roots of organic way of farming in a changed context of contemporary food demand as well as ensuring the quality of basic natural resources.

India also distinguish itself in the field of organic farming since the time of Sir Albert Howard Whose major work in this field was done at Indore in M.P. and inspired the organic agriculture movement throughout the world in 1930s. Today few group of farmer have started organic farming in a sporadic way who lack the co-ordination with notable exceptions of Maharashtra farmer in documenting and exchanging the information at appropriate forum unlike to their counterpart in Europe and United State. In order to assess the status of organic farming in India and to explore the possibility of its adoption on wider scale the Department of Agriculture and co-operation of ministry of Agriculture Government of India constituted a technical team of experts who visited different places during September, 1993 across the country including state Agricultural Universities and NGOs where the work on organic farming is going on. The team members had detail discussions with scientists, farmers and other eminent persons engaged in propagating organic farming and made the following observations', Report of Technical Team, 1993).

1. The country at present is not in a position to completely eliminate the use of chemicals especially fertilizers in view of the increasing demand for food. However, it would not be difficult and unrealistic to phase out the use of these chemicals systematically by appropriate balancing the use of optimum quantity of organic manures and biofertilizers.

2. For the control of diseases and pest the integrated pest management (IPM) approach is the right answer for indiscriminate use of pesticides.

3. State agricultural Universities are doing very little work with regard to various components of organic farming and the efforts of Extension staff of many states in propagating organic farming are negligible.

As such organic Agriculture movement in India is in infancy stage and has to go a long way to catch up the spirit of organic producers and consumers of Europe, U.S.A. Canada.

IFOAM (International Federation of Organic Agriculture Movement)

This is an international umbrella organization founded in 1972 and unites the efforts of its members to promote organic agriculture as an ecologically and socially sound and sustainable methods of food production. It includes the organic growers, processor and marketing associations as well as consumer advocate groups. The IFOAM is an essential instrument for disseminating reliable information and developing a common approach to organic agricultural and environmental issue. It represents the worldwide movement of organic agriculture and provides a platform for global exchange of information and co-operation through numerous international, continental and regional IFOAM conferences or through publication of the magazines like Ecology and farming and conference proceedings. The federation's main function is to co-ordinate the network of organic movement around the world.

The major aims and activities of IFOAM are given below

1. To exchange knowledge and expertise among the members and to inform the public about organic agriculture.

2. To represent internationally, the organic movement in parliamentary, administrative and policy making forums.

3. To set and regularly revise the international IFOAM Basic standards of organic Agriculture and food processing.

4. To make an international guarantee of organic quality a reality.

Historic concept of organic farming?

%The organic philosophy is that humans must recognize that humans can only survive and thrive if they live in harmony with the delicate balance of nature between plants, animals, earth and humans. In farming, that means using methods that conserve and enrich the soil without causing pollution, damaging wildlife or using up too many the worlds' resources. The world's shallow layer or topsoil, on which all our future food depends, contains billions of tiny live organisms in a single handful. These are essential for soil fertility. Protecting the soil and preventing its erosion are essential to our future. Organic produce is becoming increasingly popular among consumers worldwide who are concerned about the following:

Their own and their family health.

Food safety issues, to avoid diseases.

Care of the environment - especially the soil.

Overuse of fertilizers, herbicides and pesticides.

The sustainability or farming.

The increased use or genetically modified crops and animals.

Definition

The terms "organic" and "natural" in relation to food cannot he precisely define, but some meaningful generalizations can be made. "Organically grown" means that the grain, vegetables or fruits were grown without chemical fertilizers and synthetic pesticides. Some organic proponents accept the use of naturally occurring pesticides such as rotenone (from roots of the Derris plant) and pyrethrum (from the flowers of chrysanthemum species). Unprocessed phosphate rock as a source of phosphorus and greensand as a source or potassium are approved substitutes for fertilizers. The main distinction in fertility programs is the use of synthetic nitrogen fertilizers in commercial agriculture as compared to only crop residues animal and human manures and food wastes in "organic farming." "Organic" meat, milk and eggs mean that the poultry and livestock were fed "organically" grown crops and that no growth hormones or antibiotics were used. The term "health" foods usually imply that no preservatives or coloring agents have been added.

The United States Department of Agriculture (USDA) has framed a handy definition of organic farming which although it misses out some important aspects provides a description of the key practices: "Organic fanning is a production system which avoids or largely excludes the use of synthetically compounded fertilizers, pesticides, growth regulators and livestock feed additives. To the lllaxinlllm extent feasible, organic farming systems rely on crop rotations, crop residues, animal manures, legumes, green manures, off-farm organic wastes, and aspects of biological pest control to maintain soil productivity and health to supply plant nutrients and control insects, weeds and other pests".

The British Organic Fanners and Organic Growers Association has defined organic agriculture as "Organic farming seeks to create an integrated, sustainable agricultural system, relying first and foremost on ecological interactions and biological processes for crop, livestock and human nutrition and protection from pests and diseases"

Organic farming is basically a holistic management system, which pro1l1otes and improves the health of the agro-ecosystem related to biodiversity, nutrient bicycles, soil microbial and biochemical activities. Organic and bio-dynamic farming emphasizes management practices involving substantial use of organic manures, green manuring, organic pest management practices and so on. It has also come to mean that it is a system or farming that prohibits the use of artificial fertilizers and synthetic pesticides.

The **Codex Committee of World Health Organization on Food Labeling** has been guiding the approval of such products. The broad definition of organic agriculture recommended by *Codex Alilmentarius* Commission is more useful for practical purposes under Indian situation. "Organic agriculture is a holistic production management system which promotes and enhances agro-ecosystem health, including biodiversity, biological cycles and soil biological activity. It emphasized the use of management practices in prescience to the use of 'off-farm' inputs; taking into account the regional conditions require locally adapted systems. This is achieved by using where possible agronomic, biological and mechanical methods, as opposed to using any synthetic materials to fulfill any specific function within the system.

Rio-dynamic farming is an alternative variant where the chemical fertilizers are totally replaced by microbial (biological) nutrient providers such as bacteria, algae, fungi, mycorhiza, and actinomycetes. Biological Pest management of crops is undertaken by employing predators. Parasites and other plethora of natural enemies of pests, in addition to all the rest of option that help to avoid resorting to chemical pesticides. These agents could be augmented into farms or promoted through such activities that favour their flourished activities. Composting, Green manuring, crop rotations, intercropping, mixed cropping etc. as well as bird perches, trap crops promote such biological activities. Although interpreted differently by various proponents, organic farming, biological farming, regenerative farming, bio-dynamic farming. Low external input sustainable agriculture (LEISA), low input sustainable agriculture (LISA) and sustainable agriculture connote the same ideology that provide integrated efforts to maintain agro-eco systems with futurism. Organic farming could then signify all such farming practices. Shri Kapil Shah indicated that Organic farming is a philosophy for sustainable rural development and a not a technological or market option alone. This is a very important aspect because organic farming cannot be viewed in isolation as for export market but also could be a way of life in the Indian rural context. Therefore. Organic farming can it is defined as socially just, environment friendly and economically viable alternative to chemical-oriented farming. An alternative system is called 'Eco-technological farming' has often been equated to organic and bio-dynamic farming in the Indian context. This system is an effective blend to traditional practices of wisdom with appropriate modern advances of science. Integrated nutrient management (INM), integrated pest management (IPM) with optimum use of inorganic inputs is advocated.

Organic and Biodynamic Farming in Indian Perspective

Although as yet infancy, Organic Fanning IS becoming important in the agriculture sector in India, largely through the efforts of small groups of farmers. It has come out of the exploitative agriculture that has been followed by in all these years, resulting into damaging impacts on environment, human and animal health, soil and water resources. It is well known now that increased use of chemical pesticides (rather abuse) and fertilizers have created chain of problems of soil, environment and water degradation. The intensive chemical agriculture that has been followed after green revolution successes is causing heavy pollution of our food, drinking water, air, the life expectancy has improved, but the quality of life has substantially deteriorated. The rural economy is in ruins because of over-dependence of outside inputs in agriculture such as seed, fertilizers, pesticides, growth-promoting chemicals etc. It is even said that the chemical agriculture has destroyed our ability to think about the right way to go forward. Fortunately, alternatives to chemical agriculture are available in organic, biodynamic and eco-technological farming approaches. Though a small percentage of farmers are expected to take up organic farming, consumer demand for organically produced food and fiber products provide new market opportunities for farmers and farm-business around the world. In fact, Government of India has been clearly aware of the importance of organic and bio-dynamic farming approaches and the Ninth Five year plan document laid emphasis on 'Environment and sustainable agriculture', promotion of organically produced commodities, particularly in plantation crops spices and condiments. The Plan document emphasized the use of bio fertilizers, bio-control agents, and organic manures with infrastructural support. It is interesting to note that Food & Agricultural Organization on (FAO) Committee on Agriculture, during the 15th Session has discussed the topic, "Organic Agriculture" and concluded that FAO has the responsibility to give organic agriculture a legitimate place within sustainable agriculture programmes. In several developed countries, organic agriculture has come to represent a significant portion of food system (Austria, Switzerland) and many other countries such as Japan, Singapore, France, United State of America etc. arc experiencing growth rates that exceed 20 percent annually (FAO Committee on Agriculture - Agenda Item 8, pages 1-12). Many developing countries began to seize the lucrative export opportunities presented by organic agriculture (e.g., export of cotton from India, Uganda and many other countries export of Mexican coffee. organic spices etc.).

Realizing the importance of organic and bio-dynamic farming, the Planning Commission, Government of India has constituted a Working Group on the subject. The Terms of reference of the working Group are i) To review the performance of various programmes of Department of Agriculture and Co-operation (DAC) and ICAR undertaken on organic and bio-dynamic farming. ii) To assess the technical soundness of organic and bio-dynamic farming practices to provide balanced nutrition and their efficiency to exploit the full genetic potential of the recommended crop varieties. iii) To assess the techno-economic feasibility to such practices and their potentials and limitations to increase crop productivity and sustain food security of the country. iv) to suggest measures/ programmes for encouraging organic and bio-dynamic fanning practices for which these are considered feasible and viable.

In consultation with Chairman of the Working Group, all members and some of the co-opted members were contacted by providing them with available literature in the form of bulletins, books etc. and were asked to provide inputs for preparing this draft paper. M/S T. G. Kutty Menon, W. R. Deshpande, V. N. Shroff, D. Y. Ragnekar, Labangalatika Dasi, Kapil Shah etc. immediately responded by providing their learned inputs, which have been taken into consideration in preparation of the paper. Personal discussions of the Member Secretary with

Shri Kuwarji Bhai Jadhav at New Delhi on 7th August provided some additional information; since he is Chairman of the Committee for the promotion of organic farming has been engaged in preparing a report for the Ministry of Agriculture, Government of India. Internationally networked non-governmental organization, called International Federation of Organic Agriculture Movement (IFOAM), has permitted the use of certain products for soil conditioning and for pest and crop growth management. The basic rates of organic production are that natural inputs are approved and synthetic inputs are prohibited). The Hanoi declaration of IFOAM (1994) emphasizes that the Asian history of agriculture spanning into thousands of years, is in deep connection with cultural and ecological diversity. The erstwhile colonial rule as well as misdirected policies has undermined this balance. The increased rood production in certain countries also paid for the cost of degradation of traditional diversity of crops and animals. Other international movements are Commission on Sustainable Development (CSD), Society for International Development '(SID), International Union for Conservation of Nature (IUCN) and Bread for the world.

Promotion of organic farming in Maharashtra and other states of India

The seeds of commercial Indian "Organic cotton" cultivation were sown for the first time in Maharashtra in the early 1990s. Some progressive farmers, distressed by the negative effects of pesticides for insect suppression in cotton crop, reduced the chemical inputs and increased the use of organic manure, developed their own techniques to optimize resources in order to develop sustainable farm. The pioneers, in this field, arc M. V. Wankhede, S.P. Wankhede, R.S. Wankhede (from 1978 onwards) of Amaravati dist, Anantrao Subhedar. Om Prakash More and Tukaram Bhimsingh Jadhav of Yavatmal dist. (from 1990 onwards, after being introduced to the "Fukuoka" type of farming to them by Shri Dabholkar of Pune), or Shri Jain (for at least 60 last years) of Karanja-Lad who has a historical cultivation background of a few decades. They tested the 'Fukuoka' principles of farming, and stabilized their farms due to their ingenuous approaches. A team of CICR scientists visited the Yavatmal farms in 1992 crop season to analyze their package of practices. Similar efforts of promotion of organic farming have been made in many states. Efforts have been made by the NGOs to study organic farming in Gujarat, Madhya Pradesh, Kerala, Karnataka, Tamil Nadu. Agricultural Universities organized workshops, Group meetings. Seminars and Conferences on this topic drew attention of scientists to the need of research in this area. The use of bio fertilizers, bio pesticides, vermin-compost, Farmyard manure, green manure, crop residues have been based on long experimentation. In fact, a number of farmers, NGOs and even some Universities/ Institutions are practicing organic farming, using traditional sources and methods of nutrient supplies to the crops and non-chemical forms of plant protection measures with varying degree of success. However, the technology adopted and methods followed are not well- documented. There has been a good suggestion from Shri Kapil Shah who emphatically proposed that the Universities and institutions that are undertaking organic farming trials should not only go about INM and IPM. It is essential that the philosophy of organic farming is percolated to them properly. It is for the scientists. Basically to evolve separate protocols for the conduct of organic farming trials. A distinction is necessary at this stage between the trials or organic farming and biodynamic farming including the eco-technological approach wherein the organic inputs approved by the certifying agencies arc tested in replicated or large-plot trials instead of combining INM and IPM.

An ad-hoc field experiment was begun in 1991 at CICR farm in 600 sq. m. plot to marry the agronomic techniques to reduce the fertilizer component while the pest management was exclusively through biological means. LRA 5166 was the variety that was invading central zone from southern zone and was chosen as the cultivar for the experiment, although it was known to be more susceptible to Jassids. Parallel plots of similar dimensions were kept with the recommended practice of fertilizers and insecticides. A third plots received 50% of the inputs of both the other plots. Another set of 3 plots was cultivated with G. arboretum *(Desi)* cotton variety. *AKA* 5 and later with AKA 8401. This was planned as a long-term trial.

The initiatives taken by Bombay Burmah Trading Corporation (BBTC) to convert Singampatti group of estates in southern India to market organic tea internationally is the first of the few efforts for commercial organic agriculture. A number of organizations such as The Ecological Development Society in Pondicherry. Institute for Integrated Rural Development at Aurangabad, The Society for Equitable Voluntary Actions (SEVA) in West Bengal, The Indian Agency for Organic Agriculture (IAOA), Peekay Tree Crops Development Foundation (PTCDF) at Cochin undertake training of personnel towards organic farming. All India Federation of Organic Farmers promotes organic farming in the country.

It has been realized that India can play a major role in International organic spices market. The Spices Board (K.S. Nair in Business News) has emphasized the expansion of Indian spice market because it is fell that the spices growers have a natural 9 advantage in terms of large tracts of land in the tribal belts of Orissa, north-eastern states, Nilgiri hills and Andaman & Nicobar Islands where traditional practices are still in vogue. The Board has already launched the Schemes to assist the organic spice growers by publishing national standards for organic spice production. It has received approval from the National Standards Committee of IFOAM. During 1997-98, India has exported 32.01 tonnes of different organic spices as against 25.32 tonnes in previous years.

During the early part of last decade, the concept of Sustainable Agriculture and Rural Development (SARD) was introduced in one FAO conference held in Netherlands to attain food security, employment and income generation in villages and natural resource conservation leading to environmental protection.

The intense development and progress of organic farming in our country also synchronized with this resolution as well as similar resolve of United Nations Development Programme (UNDP, 1994).

Besides these efforts of ICAR and other NGOs in promoting organic and biodynamic farming, a real government support either in the form of subsidizing the organic inputs or production promotion schemes of FYM, NADEP and other biological composting methods, *AMRUTPANI*, green manuring, recycling of farm wastes using earthworms or *Trichoderma* spp., botanical pesticides, biocontrol agents etc. has not been visible. There are sporadic efforts on the part of the State governments to organic and biodynamic farming. For example, the Maharashtra Cotton Marketing Federation has purchased organic cotton separately and helped the growers in exporting to EU regulations and other countries.

If you are considering changing to organic farming, you may have a few critics who will repeat the theory that you will not survive financially, you're profitably and production will have to fall. Be clear on these points. Organic farming is NOT:

Farming by neglect.

Losing money.

The road to bankruptcy

Letting your farm run down and look like a wilderness.

Letting pastures and crops fail through low fertility and disease.

Letting animals perform poorly because of poor feeding and disease.

Letting animals suffer, and breaking the law (Animal Welfare Act 1999).

AI ways remember that to be a good organic farmer, you'll have to be an above-average conventional farmer. The farmer who is struggling to be profitable in the conventional system will not adapt well to organic farming, hence they tend to balance the system for their failure, rather than themselves. Therefore, a balance must be struck between the spiritual and material side of life.

Why we have to follow organic farming?

Food and nutritional security is important to any country and India is not an exception to this phenomenon. Agriculture has a very important place in Indian economy and society to meet the food grains requirement of the ever-increasing population, which has already crossed a billion. The land to man ratio is declining. The food grain production in the country has increased steadily from 51 MT in 19520-1951 to 212 MT during 2001-2002 due to scientific management of inputs viz seed, fertilizer, pesticides and water. Food security for our ever-growing population has been made it possible by raising high-yielding varieties of crops, adopting judicious use of irrigation and integrated nutrients and pest management practices. Continuous increase in the level of input use has resulted in sustaining self-sufficiency in the production of food grains. Although it has often yielded the desired results but ironically even stagnating trends in crop productivity are observed recently in several crops. This paradoxical situation is mainly attributed to a steady decline in the efficiency of 1110st inputs used in crop production. Apart from low yields, poor efficiency of fertilizers and agricultural chemicals also adversely affect soil health and quality of food and environment. The increased food grain production relates to fertilizer consumption. The food grain n production and fertilizer consumption, however, declined to 182.6 MT and 16.1 MT respectively due to drought in 2002-2003. This has again brought forward the need for efficient utilization of all resources by way of in-situ soil and water conservation, increasing moisture use efficiency and integrated use of fertilizers, manures, bio-fertilizers and city wastes for sustainable agriculture. Despite all these benefits from fertilizers, the eco-friendly environment people blamed for causing irreversible damage to ecosystem from global warming lo hypoxia to deformed frogs and their environmental problems. The scientists have also realized that the "Green revolution" with high input use has reached a plateau and is now sustained with diminishing returns of dividends. The nutrient use and factor productivity is on decline. Thus this agro-chemical based technology is no more sustainable and has given rise to second generation problems of "green revolution" like land degradation, pesticide residues in farm produce, water and atmospheric pollution. Besides, there is a change in soil reaction, decrease in microbial activity, reduction in organic matter and development of nutrient imbalances / deficiencies and declining soil fertility. This needs to be corrected by adopting alternate technology of farming, a technology that should be eco-friendly, farmers friendly and economical. It should help in perceiving traditional bio-diversity and knowledge.

A new strategy of living with the nature and nurturing for sustainable high productivity should be evolved. The rational use of manures and fertilizers but not exclusive use of either helped in improving -productivity under prevailing tropical condition. However, it is important that the transition from conventional to organic farming should take place gradually and effectively so that food production does not suffer by adopting alternate farming system. Organic agriculture shows us the way to effectively use the available natural resources for the benefit of the mankind. Organic farming is an alternative agriculture production system, which largely excludes the use of synthetically compound fertilizers, pesticides and growth regulating hormones. Organic agriculture should meet the dual challenges of sustaining the food productivity while causing minimum damage to the agro ecosystem and environment. Total organic farming may be the most desirable proposition but is not feasible due to inadequacy of organic manures, slow nutrient releasing capacity and poor source of nutrients in modern commercial agriculture and cannot sustain at high levels of production in prevailing semi-arid or tropical climatic conditions. Organic materials can differ widely in their properties and characteristics. The plant nutrient contents of most organic materials are generally much lower than those supplied by commercially available chemical fertilizers. The micronutrient content of crop residues can range from 0.7 to 2.5 percent for nitrogen, 0.07 to 0.2 percent for phosphorus and 0.9 to 1.9 percent for potassium; animal manures can range from 1.7 to 4 percent for nitrogen, 0.5 to 2.3 percent phosphorus, and 1.5 to 2.9 percent for potassium (USDA, 1978); while sewage range from 3 to 7 percent for nitrogen, I to 3 percent for phosphorus and 0.2 to 3 percent for potassium (Hornick et al. 1984) It has been reported that out of 23.17 and 27.83 MT of total wastes, 13.5 and 16.03 MT could be made available for utilization with fertilizer equivalent of 6.75 and 8.13 MT by 201 I and 2025, respectively (Amrutsagar and Shinde, 2004). About 28-30MT of nutrients are removed from the soil while not more than 18.20 MT of nutrients are added through all sources living a gap of 10 MT. According to recent estimation of planning commission of India, to produce 300 MT of food grains. 40-45MT of nutrients will be needed by 2025 (Basavannepa and Biradar, 2003). Paroda (1997) reported that none of the sources of plant nutrients is adequate. Organic sources can best be supplemented with fertilizers, which will increase the efficiency of nutrients.

Total organic farming would be possible only under substance farming practiced few decades back on soils that are already at a relatively high level of fertility and productivity in our country. It might be possible for horticultural farming where consumers are ready to pay extra premium for the "green" products to compensate for their lower productivity.' There are challenges for adoption of this technology, as it requires scientific explanation for organic produce, quality of its inputs, consumer's awareness and formulation of standards for inputs and produce, certification of farm produce and processing etc. Organic agriculture is more of a process than an end product. Thus, organic production refers to organically grown crops, which are not exposed to any chemicals right from the stage of seed production to the final post harvest handling and processing. There are many restrictions on sourcing of external input even if they are organic (urea fertilizer) It is based on the recycling of natural organic matter, crop rotation green manuring, animal manures etc. for improving soil fertility. Therefore to meet entire nutrient requirement of crop from alternate sources is a greater challenge. Supplemental nutrition of organic soil via off farm manure and soil amendment as incorporation of composts, manures (bulky and concentrated), green manure, biological nitrogen fixation (Azotobacter, Rhizobium, Azopirillum), PSB, BGA, Azolla, YAM, bio-fertilizers, use of legumes in crop rotation, animal byproduct as blood meal, fish meal, horn meal, non edible oil cakes, mushroom compost, soil mineral sources as powdered rock phosphate, kainite, langbeinite, natural mined K_2SO_4, agricultural lime stone and gypsum, calcified seaweed, epsom salt (hydrated magnesium sulphate) kelp weed, rock powders etc. be followed to meet all essential nutrients for plant growth and development. However, NPK fertilizers, quick lime, slaked lime, excessive micronutrient salts, synthetic growth promoters, adjutants, wetting agents are prohibited. Any imbalances, deficiencies or excesses which are identified must be addressed. The producer shall establish a long -term soil quality-monitoring program. Cropping system promote, synergism and complimentary for increased productively (Narwal, 1994, lampkin, 1992). Cultural crop rotation and mechanical practices for effective weed management and mechanical. An integrated organic management (10M) system is therefore advocated to achieve the common goal of present day agriculture. Organic sources of nutrients to sustain soil fertility at current level of production and biological and cultural methods of pest control are the two important pi liars of successful organic farming. It is known by different names in different countries as Organic farming (UK and USA), biological farming (Europe), and regenerative farming sustainable farming. Biodynamic farming (Rudolf Steiner- Austria), nature farming, perma-culture, alternate agriculture, ecological agriculture, Integrated Intensive Farming System (IIFS), Low External Input Supply Agriculture (LEISA) etc.

The distortion in soil fertility and deterioration in soil health is due to improper and indiscriminate use of irrigation and certain NPK fertilizers. The lopsided fertilization practices have led to imbalances in the status and availability of plant nutrients, which can be corrected only with proper manure fertilizer schedule based on soil fertility evaluation. Integrated soil fertility management using manures, mineral fertilizers, bio--fertilizers, green manuring, crop rotation, mixed cropping with legumes and residue management besides biological pest management will facilitate restoration, improvement and maintenance of soil fertility, sustaining productivity and soil quality on long-term basis (Swarup et al, 2001). This will insure and ensure agricultural production at high levels with good quality produce as well. It will also safeguard the environment and natural resources. Organic farming is one step ahead of INM and IPM. It prohibits use of synthetic chemicals either as pesticides or fertilizers. Every farmer experiences that good fortune i.e. good weather or absence of pests and diseases etc. season after season is the most essential requirement for a successful crop production. Organic farming principles are based on the LEISA (Low External Input Sustainable Agriculture). Organic food and fiber production is holistic concept.' The goals and objectives are to:

1. Protect the environment, minimize soil degradation and erosion, decrease pollution, optimize biological productivity and promote a sound state of health.

2. Replenish and maintain long-term soil fertility by optimizing condition for biological activity within the soil.

3. Maintain diversity within and surrounding the enterprise and protect and enhance the biological diversity of native plants and wildlife.

4. Recycle materials and resources to the greatest extent possible within the enterprise.

5. Provide alternative care that promotes the health and behavioral needs

of live stock.

6. Produce sufficient quantity of food of high quality.

7. The main objective of organic soil management programme in cultivated system shall be the establishment and maintenance of a fertile, friable soil using practices that maintain or increase soil humus levels, promote an optimum balance and supply nutrients, stimulate biological activity within the soil and maintain good soil tilth. Organic agriculture is gaining momentum all over the world especially among small farm holding, as it enhances farmer's ability to live in harmony with nature and to derive economic benefits. Small and marginal farmers over the centuries have experimented, innovated, adopted and standardized agricultural practices suited to their agro- climatic regions and socio-economic conditions. Some of the practices. one as follows:-

1. Use indigenous seed having high potential but grown on organics rather than genetically engineered seed.

2. Soil and water conservation measures are adopted using mechanical and biological barriers, cover crops, mulch etc.

3. Enhance soil biological activity with respect to soil micro flora for improving soil health, rehabilitation and nutrient uptake by plants, bio-degradation processes reducing hazardous waste and control of pests through natural bio--control.

4. Development of alternative management to take care of microclimate, drainage and organic matter status.

5. Identification of suitable micro flora for speedy decomposition of on farm wastes into organic manure and varmi-compost.

6. Nutrient requirement is met on the basis of assimilation of nutrient by plant rather than application to the soil, as the recovery through organics is high over inorganic nutrients.

7. Use compost and liquid manures, green manure, bio-fertilizers. The other options for nutrient recycling are cow pat pit, BD-500 spraying of vermiwash, BD liquid manure, Amrut pani, Panchamrut, Agnihotra etc.

8. Exclude plant protection chemicals for pests and disease control. Emphasis on use of Bio-control for pests by using natural parasites, predators and beneficial fungi, bacteria and viruses is effective and devoid of problems associated with chemical pesticides.

9. A number of plant extracts from herbs, weeds, vegetation which are unpalatable to animals have potential for plant protection.

10. Use companion plant as inter cropping or in crop rotation system.

11. Diverse species like fishes, honeybees, mixed in terms of reduced external inputs, increased production and sustainability in production system.

12. For diseases control, copper and sulphur-based compounds are recommended, in addition to use of microbes, viz. *Trichoderma, Aspergillus* can be used in organic production.

The beneficial effects of organic on soil physical properties such as soil aggregation, increasing water retaining and releasing properties of soil, optimum soil-air-water relations in the rhizosphere, lowering soil compaction by decreasing bulk density and soil crusting, increasing infiltration and hydraulic conductivity, enhancing biological properties of soils, minimizing pest built up in soil and developing immunity against pest in plant prevent soil degradation and deterioration by conserving resources through minimizing soil and water erosion, prevent nutrient losses through runoff and leaching, maintain tilth, fertility and productivity agricultural soil. Restoration of soil fertility by recycling residues and improving nutrient availability, providing all essential nutrients for plant growth, etc. is well known and established. As such there is no loss of productivity by it adoption contrary to this, it gives better production under protected cultivation practices. It improves the quality of produces. It is cheaper, labour intensive and provides opportunities to increase rural employment. The diversity of crops and livestock gives the farmer flexibility and a diverse income. The energy consumption is for less than conventional farming moreover energy efficiency is high. This higher level of organic stability provides a distinct advantage in the initial reclamation of marginal soils because it imparts beneficial and long-term residual improvement 01 soil physical properties unless the physical nature of soils is improved first, the plant use efficiency of nutrients, whether from organic amendments or chemical fertilizers will be unacceptably low (Parr et al., 1986). The primary goal of organic agriculture to optimize the health and productivity of interdependent communities of soil life, plants, animals and people. Thus organic farming is a self-sustaining system of agriculture and offers a viable potential alternative and provides a balanced environment. This helps in maintaining soil fertility and controls pests and diseases by the enhancement of natural processes and cycles with moderate inputs of energy and resources while maintaining an optimum productivity level with maximum harmony with the environment. Organic agriculture is a holistic production management system, which promotes and enhances agro ecosystem heath, including biodiversity, biological cycles and soil biological activity. It emphasizes the use of management practices in preference to the use of off-farm inputs. This is accomplished by using agronomic, biological and mechanical methods. It helps in maintaining soil heath and sustains production with carrying capacity of ecosystem.

The different problems or issues related to organic farming

Organic agriculture is a holistic approach based upon a set of processes that leads to sustainable ecosystem, safe and nutritive food, animal welfare and social justice. It is based on minimizing the use of external inputs and avoiding the use of synthetic fertilizers and pesticides. It is the process of farming system employing management practices, which seek to nurture ecosystem, achieve sustainable productivity and provide weed, pest and disease control through a diverse mix of mutually dependent life forms. Thus organic agriculture encourages a balanced host/ predator relationship through augmentation of beneficial insect populations, biological and cultural pest control, recycling of plant and animal residues. Soil fertility is maintained and enhanced in a sustainable manner by a system, which optimizes soil biological activity and the physical and mineral nature of soil as the means to provide a balanced nutrient supply for plant and animal life as well as to conserve soil resources with the recycling of plant nutrients as an essential part of the fertilizing strategy. Pest and disease management is attained by means of crop selection, rotation, water management and tillage (Smitra Das and Biswas, 2002).

World market survey conducted during 2001 reveals that organic agriculture is being practiced in about 17 M. ha of which 7.7 M ha in Australia, 2.8 M. ha in Argentina, about 4.2 M. ha in European countries. It is expected that equal area will be brought under organic farming in near future. Simultaneously, the organic market has also grown from $ 4.0 Billion (1992) to $ 21 Billion in 2002. Even European countries have earmarked about 5.3% of the total budget for promotion of organic agriculture. Moreover, U.K. Switzerland, Denmark, Sweden countries are planning to achieve the organic growth rate to the range of about 30-40% by 2005. India has just started achieving 75000 ha area under organic agriculture covering 10,000 ha tea, and medicinal plants each and 12,000 ha cotton (Chhonkar P.K. 20(3). There is no doubt about its advantages, scope and limitations in our country. Codex Alimentarius FAO/WHO indicated that food described using the term organic area product of organic farming, which is a system of farm design and management practices that seek to create eco-system, which achieve sustainable productivity and provide weed and pest control through a diverse mix of mutually dependent life forms, recycling plant and animal residue cropping pattern and crop rotation, water management tillage and cultivation. Soil fertility is maintained and enhanced by a system, which optimizes soil biological activities as the means to provide balanced nutrient supply for plant growth. These practices will also conserve soil resources and protect the environment. There are specific International standards e.g. Codex Alimentarius Commission Guidelines. Moreover, Regional, National and Certification standards have been framed and are available to organic farmers. Different regions in the world are evolving regional standards for organic agriculture (Pant R.K., 2004). European Union has already evolved their own standards and common legislation on organic production (E. U. Council Regulation EEC No. 2092/91). It is a matter of concern that agro-synthetic pesticides are not used in organic agriculture. Non-application of mineral nitrogen fertilizers in organic farming are to be strictly observed. The soil life is enlivened and soil fertility is sustained by providing O.M. and humus.

Definition of organic farming

Organic farming is production system which avoid or largely excludes the use of chemical fertilizers, pesticides and growth regulators. It also depends upon crop rotation with leguminous crop, addition of crop residues, animal manure's, green manuring, bio-fertilizers and bio-pesticides etc. United States Department of Agriculture (USDA) has defined the organic farming as a production system, which avoids or largely excludes the use of synthetically compounded fertilizers, pesticides, growth regulators and livestock additives. Naturally organic farming systems rely on crop rotations, crop residue, animal manures, legumes, green manures. Off-farm organic wastes and the aspects of biological control to maintain soil productivity and tilth to supply plant nutrients and to control insects, weeds and other pests. The concept of the soil as a living system that develops the activities of beneficial organisms is central to this definition (Kalloo, G. 2004).

Why organic farming

To avoid adverse effect of farm practices

Food security and quality

To maintain sustainability in production and productivity

Environmental security

Health Security of human and animals Maximum utilization of organo materials Self sufficiency

To have good participation with our produce 111 international market (WTO).

Resources of organic farming

- Crop residues
- Livestock and human excreta

  Green manuring Bio-fertilizers

  Utilization of Agro waste materials Natural green biomasses

  Industrial wastes like press mud, spent wash vermicompost etc.

Components of organic farming

The components of organic farming are as

Organic manures

Organic materials such as farm yard manure, biogas slurry, compost + straw of other crop residues, biofertilizers, green manures and cover crops can substitute for inorganic fertilizers to maintain environmental quality. In addition, the organic farmers can also use sea weeds and fish manures and some permitted fertilizers like basic-slag and rock phosphate.

Non-chemicals weed control measures

Compared to conventional farmers, the organic farmers use more of mechanical cultivation of row crops to reduce the weed menace. No herbicides are applied as they lead to environmental pollution.

Biological pest management

The control of insect pests and pathogens is one of the most challenging jobs in tropical and sub-tropical agriculture. Here again non-chemicals, biological pest management is encouraged. The conservation of natural enemies of pests is important for minimizing the use of chemicals, pesticides and for avoiding multiplication of insecticide-resistant pests. Botanical pesticides such as those derived from neem could be used. Selective microbial pesticides offer particular promise, of the strains of *Bacillus thuringinensis* is an example.

Management of organic farming

Management of organic farming system involves

1. Organization of crop and livestock production and the management of farm resources in such way that it hormonizes rather than conflicts with natural systems.

2. Achievements of a closed cycle to the greatest extent possible between soil, plants, animals and people and an avoidance of environmental pollution.

3. Maintenance of soil fertility for optimum production, relying primarily on pollution.

4. Reduction of pest and disease incidence through a carefully designed farm rotation and enterprise structure: use of resistant varieties; the encouragement of beneficial pest predators and the use of other biological pest control techniques.

5. Use of forms of animal which respect the welfare and behavioural needs

of farm livestock.

6. Use of appropriate farm machinery and cultivation techniques which reduces non-renewable resource livestock.

7. Enrichment of the environment in such a way that wild life flourishes and it is enjoyable for people both working within the system and viewing it from outside.

Publicity, Awareness and Education of Farmer

(a) Promotion of organic farming involved educating the farmers about its benefits. The State Governments should take up awareness campaigns and use media, would first be necessary to familiarize its own officers/staff with the concept and practices of organic farming. Many NGOs and other agencies have become involved in promotion of organic farming. Their assistance/ cooperation can be sought to make organic farming popular.

(b) The State Governments may take assistance of SAUs/ KVKs in developing soil and area specific package of practices, strictly in conformity with the principles of organic farming, for adoption by the farmers in the area. This information could be publicized through all possible means.

Regulatory mechanism

Adoption of organic agriculture necessarily involves a sequence of steps that need to be followed by the growers and verified by certification and inspection agencies. This is necessary to ensure that the consumer is not duped and genuine organic cultivator is not put to disadvantage. Important steps which State Governments may take are the following.

(a) Formation of organic farmers group

The farmers with similar farming and production system should be persuaded to form a farmer's group, preferably in the same village with continuous area. Number of farmers in a group may vary depending upon local dictionary. Formations of such group will facilitate inspection and certification, monitoring and supervision, etc.

(b) Registration of farmers group with district authorities

The registration of farmers groups taking up organic cultivation may be done by the State Govt. designated district level authority. This authority could be 'constituted under Department of Agriculture or Horticulture. For registration, only intimation may be prescribed, receipt of which may be acknowledged. This may constitute registration. The State Government should prescribe a proforma for giving intimation. This should include information about individual members, plot number, area & crops to be covered etc. Information contained in the intimations received should be compiled periodically and made available to the Department of Agriculture & Cooperation.

(c) Documentation of individual farms / farmers records

Documentation being one of the prerequisites for the certification of organic product, certain documents is to be maintained for individual farms and farmers. These documents provided by the certification agencies will be maintained and updated periodically both at the level of individual farmers and at the group level. An educated farmer from the group after training may be assigned the responsibility of maintenance of required documents for the entire farmers' group to the satisfaction of certification agencies.

In case such a person is not available, Service Providers registered with State Government may provide this service at nominal cost.

Service Providers

The State Governments may register Service Providers. These may be KVKs/SAUs/Agri. clinics/Farmers Groups/ATMAS/NGOs/Private Entrepreneurs/Central Agencies, etc. The State Government should select area specific Service Providers, based on their credibility and expertise. To begin with SAUs, KVKs, Agri-clinics, ATMA, Central Agencies and reputed NGOs already active in the field of organic farming may be approved as Service Providers. Later, other private sector trained persons can be registered as Service Providers. The service providers will help in the documentation, development of package of practices and providing day-to-day advice to the farmers. Service Providers being local agencies. Well-versed with the agronomic practices, availability of organic inputs and the technology, will provide necessary advice to the farmers group and will serve as real friend in need.

The, State Governments may fix nominal fee to be charged by private Service Providers from individual farmers for the facilities provided. Some Service Providers could also become input supplier on their own as commercial venture . However, a Service Providers cannot become inspection agent or a certification agency. The State Government may prescribe minimum qualification necessary for a private Service Providers. It is suggested that, he should be matriculation with Training in any of the recognized institution. The Training should be for a period of one month.

(*e*) Accreditation Agencies

Presently, the existing 6 Accreditation agencies approved by Ministry of Commerce are authorized.

Agricultural and Processed Food Products Export.

Development Authority (APEDA) II. Coffee Board.

Spice board.

Tea Board.

Coconut Development Board VI. Cocoa & Cashewnut Board.

(*f*) Certificate and Inspection Agencies

Since the organic farming is a 'Process Certification' and not 'Product Certificate', the role of certification agencies is most critical. The Certification Agency has to be impartial and a Non Government Agency. Its accreditation by an authorized Accreditation Agency is mandatory. Presently, there are only 4 certification agencies accredited by APEDA.

The certification agency may appoint one or more inspection agents by entering into a written contract, specifying the terms and conditions of their agreement. It is basically the responsibility of certification agencies to ensure strict compliance of national standards of organic farming. They prescribe specific documents to be maintained at the level of farmers/ farmers group. They also prescribe minimum conversion period after inspection of farm and other details. They are also authorized to issue necessary certificate of organic production to the farmers.

The state Governments are advised to encourage formation of local Certification Agencies fulfilling the requirements. Details can be obtained from Ministry of Commerce / APEDA. The Inspection agencies fulfilling requisite criteria may approach the accreditation agencies. Only after the issuance of accreditation certificate by the accreditation agencies, the concerned agencies can function as a certification agency.

(*g*) Periodic inspection of organic farms

The certification agencies either themselves may inspect the records of the organic farmers groups or may appoint inspection agencies to ensure proper compliance of the process of organic agriculture by the farmers. The certificate agencies, if need arise, may get the samples of soil, water, organic inputs, pesticides etc. and get them tested in their own laboratories or recognized laboratories to decide about the conversion period, as also to ensure the adoption of National Standards of Organic Farming.

(*h*) Formulation of National Standards of Organic Farming

The standards already prescribed by the Ministry of Commerce continue to be valid standards for both domestic and export markets.

(*i*) Conversion period

The time between the start of organic management and certification of crops is known as conversion period. It may vary from one kl three years, depending upon the current usage of chemical fertilizers and pesticides and past use of land. This will be determined by certificate agencies.

(*j*) Crop Production

All seed and plant material used should be from same farm and, as far as possible, be adopted to local soil and climatic conditions. Use of genetically engineered seeds, trans-genetic plants / plant material is strictly prohibited.

(*k*) Manurial policy

To the extent possible, compost prepared from on-farm organic residues/ resources needs to be used for maintaining soil fertility, humus content and biological activity. The accumulation of heavy metals and other pollutants should be avoided. The dose of highly rich nitrogenous organic material like dried blood, poultry waste, compost and slaughterhouse waste etc. should be decided by certification agency to have minimum adverse effect on the quality of crop and environment, particularly, ground and surface water.' The uses of synthetic fertilizers are strictly prohibited. An indicative list of permissible crop protectants and growth regulators is given in Annexure-II.

4. MONITORING

The State Governments are requested to send details of farmers groups registered with the State Department of Agriculture & Cooperation.

Standards and type of standards

International Standards for organic agriculture approved by International bodies and recognized by legal authorities e.g. Codex Alimentarius Commission,' Guidelines for the Production, Processing, Labeling and Marketing, of Organically/Biologically Produced Crops -IFOAM Basic Standards.

Regional Standards: Different regions in the world are evolving regional standards for organic agriculture.

National Standards: Basic organic standards prepared by respective countries.

Certification Standards: Detailed standards as compared with general framework of basic standards set by National Standards. These are available with certification agencies.

Importance of Regional-National Standards

Any Government, that wants to be listed under 3rd country list of European Commission for exporting organic products into Europe, it is necessary to have approved National Standards. Help to integrate tested technologies in traditional forms of agriculture production with the main stream of technologies for organic agricultural practices.

Help to promote extension and research programmes in a systematized manner.

Rationale for Regional Standards

To facilitate partnership between nations for trade, etc.

To enable mutual cooperation between alternative marketing channels working in different countries.

To facilitate accreditation programmes and certifying equivalency of certification programmes in different countries in the region.

- To achieve continuous exchange and update of information and technology.

Organic Farming and Non-application of agro-synthetic pesticides

- Agro-synthetic pesticides are not used in organic agriculture, but a simple non-application of agro chemicals does not mean that a farm is organic.
- A complete conversion of the farm **in** compliance with an organic management plan is essential for organic production.
- The mannural policy of the farm should comprise of taking up farm's own organic matter into the nutrient cycle.
- Preventive measures of weed and pest management should be incorporated.

Pesticide residues in organic produce

- Synthetic agro-chemicals are not used in organic food production; however the substances from 'permitted lists' can be used for plant protection and fertilization.
- No residues from active usage of organo-synthetic pesticides can be present in organic products.

When traces of chemicals are detected in organic food, the inspection body is responsible to determine the cause and the manner of contamination.

Organic food tastes better and is of superior quality

The traditional belief that organic manure promotes quality while mineral fertilizers promote quan-tity was shown to be over-simplistic by Schuphan (1974) on the basis of trials conducted for over a decade. Regardless of whether the nutrients are from organic or inorganic source, plants absorb the same in the form of inorganic ions: ammonium, nitrate, phosphate, potassium etc. Sensors in plant roots, if any, to distinguish between nutrient ions coming from organic or inorganic source have to be still discov-ered. Once absorbed the nutrients are resynthesized into compounds which determine the quality of pro-duce *e.g.* flavour, shelf-life etc., which is the func-tion of genetic makeup of the plants (variety). There is no scientific evidence presented as yet to show that organically produced food is of better quality and taste, and use of chemical fertilizers deteriorates it. The better taste of the organically grown food is of psychological in nature and could be attributed to 'Placebo effect' widely used in drug testing, where harmless sugar pills administered to control groups are known to cure patients of their imaginary ail-ments, when told of its novelty and wonderful thera-peutic properties. More someone pays for it faster is the cure: a clear case of 'mind over matter'.

Organic food is more nutritious and safer

There is a general perception in public minds that organically grown food is more nutritious, healthy and safe. There are no consistent and valid reports of differences in the mineral contents of organic and conventional food. However, N applications generally improve both the protein and bread making quality. There are many factors, environmental and cultural, that influence the nutritional composition of the pro-duce. It is at best confusing to give credit for these. Changes to organic cultivation. There is no difference between the protein content and other qual-ity parameters such as vitamins, nutraceuticals and trace minerals of conventionally and organically grown crops which at best could be linked to the varietal characteristics. The genetically modified 'yellow rice' which was in the news recently owing to its higher vitamin A content over the traditional varieties will continue to have its superior nutritive value irrespec-tive of organic or inorganic fertilization. In the field of plant nutrition the cry of 'only natural has no justification or scientific basis (Woese et al. 1997). The altitude that organic foods are safe and healthy is based on misconception that hazards in food are mainly derived from agro-chemical additives. In fact, the microbes and not chemicals are the major sources of the food-borne diseases such as typhoid, gastroen-teritis, dysentery, cystecurcosis, *etc.* Animal waste can be an effective nutrient source but the pathogen risk must be seriously considered. Animal wastes contain intestinal bacteria, many of which may present substantial human health threats. Land application of manure is particularly associated with *Salmonella, Es-cherichia coli* and *Taenia soleum* which can contami-nate the soil. These pathogens are known to survive in soil for a long period. They may be carried on edible plant parts coming in direct contact with soil and get into the food chain. They may also be intro-duced into shallow surface waters as well as ground water polluting potable water supply (Mikkelsen and Gilliam 1995).

Organic farming is eco-friendly

It is advocated that organic farming is eco-friendly. It keeps the soils healthy and does not pollute environment. It is well known that nitrate is the main end product of manure decomposition and it is continuously released from organic matter under-going decomposition. Since nitrate release is not syn-chronized with either crop demand or its uptake, it therefore tends to accumulate in excessive amounts in soil and poses environmental risk.

Nitrate thus formed without being taken up by plants may leach polluting ground water or may deni-trify polluting atmosphere. The ions irrespective of their origin whether from organic or inorganic source will behave similarly? There is no evidence that NO_3 ions from organic sources are less mobile or have lower denitrification potential than from chemical fer-tilizers. Trace elements and heavy metal concentra-tions in animal wastes (manures) and sewage slud-ges can be at times very high and often exceed concentrations normally found in chemical fer-tilizers. Field application of such organic manures which have to be applied in very high quantities in order to meet the requirement of major plant nutrients may lead to heavy metal accumulation in soil polluting arable land. These will find way into edible plant parts and will get into the food chain becoming a health hazard. From Permanent Manurial Experiments con-ducted at Coimbatore (Tamil Nadu) it has been ob-served that continuous addition of cattle manure for sixty-eight years resulted in substantial build-up of Zn, Mn, Cu, and Cd in the soil (Kurumthottical, 1995). Increased emissions of green house gases are thought to accentuate the potential for global warm-ing and climate change. Carbon dioxide, methane and nitrous oxide are the three most important green house gases associated with agriculture. Carbon dioxide emissions come, besides from burning of fossil fuels, from decomposition of soil organic matter and crop residues. Methane originates predominantly from cattle manure - an integral part of organic farming. Nitrates formed in excess of the crop demand may further denitrify to yield nitrous oxide, its emission is further increased as denitrification is positively correlated with microbial biomass and organic matter which tend to increase on manuring basis.

How Indian farmers maintaining soil fertility using different management practices?

The Maharashtra is the third largest state in India. Major soil groups of

Maharashtra are eight and there are nine different agro-climatic zones. The major crops grown arc sorghum, groundnut, wheat, pearl millet, sugarcane, grape, cotton rice and pulses, groundnut - wheat sugarcane wheat are the common crop sequences followed. The different indigenous agro-climatic organic manures are used viz. FYM, poultry manure, crop residue of lopping, neem cake coated urea, spent wash, vermicompost, press mud cake, flyash and Celrich. The treatment combination receiving the recommended dose of fertilizers and 3t/ha or poultry manure to ground nut recorded maximum and significant increased in dry matter over rest of the treatments. The per cent increase in yield over control varies from 9.3 to 19.6 under different treatments of Jalshakti and *rhizobium,* highest being under *rhizopbium* seed treatment followed by 50 kg N/ha was applied through crop residue + leucaena lopping. The use of organic source like FYM shows 25 per cent saving or nitrogen. FYM (g) 25 % dose was found best among the tested organic sources. The effect of 75 % of gypsum + neem coated urea in saline sodic soil increases the $NH_4 - N$ and $NO_3 - N$ in a soil. The compost prepared from spent wash and press mud and spent wash solids are using by the farmers in Maharashtra. Vermicompost was found to be a better source of organic manure than FYM. The possibility of substituting 25 and 50 kg N/ha of recommended dose of nitrogen by FYM and vermicompost respectively. The beneficial effects of use or crop residue with leucaena loppings or urea in soil for release of mineral nitrogen were noticed. The application of flyash upto 10 t/ha was more effective in terms or yield of summer green gram and also improved the physico-chemical properties of soil. The application of Celrich @ 3.75 t/ha to groundnut obtained a highest yield.

Maharashtra is the third largest State in India and occupies a total Geographical area of 30.77 million hectares spread over 33 districts. About 59.2 per cent area is under cultivation and 20.8 per cent under forest. The irrigated area is 15.8 per cent of the crop in the State.

Soil

Major soil groups of Maharashtra and their classes according to USDA system are given as below:

1. Shallow black soils.

2. Medium black soils

3. Deep black soils

4. Lateritic soils

5. Coastal alluvial soils

6. Saline alkali soils

7. Mixed red and black soils

8. Red loamy and red and yellow soils

Agro-climatic zones of Maharashtra

Maharashtra is divided in to nine agro-climatic zones viz.

1. Very high rainfall zone with lateritic soils

2. Very high rainfall zone with non lateritic soils

3. Ghat zone

4. Transition -I with red to reddish brown soils

5. Transition -2 grayish black soils

6. Scarcity zone

7. Assured rainfall zone with *kharif* crops.

8. Moderate to moderately high rainfall zone with soils formed from trap.

9. High rainfall zone with soils formed from parent material

Crops

Major crops grown in Maharashtra are sorghum, groundnut, wheal, pearl millet, sugarcane, grape, cotton, rice, and pulses, groundnut-wheat, sorghum - gram, pearl millet - wheat, cotton - wheat, sugarcane - wheat are the common crop sequences followed.

1. Farm yard manure

In Maharashtra, almost all farmers from 33 district using FYM. Good quality farmyard manure is perhaps the most valuable organic matter applied to a soil. It is the most commonly used organic manure in Maharashtra. It consists of mixture of cattle dung, the bedding used in the stable and of any remnants of straw and plant stalks fed to cattle. Though its crop increasing value has been recognized from time immemorial, more than 50 per cent of the cattle dung produced in the state today is burnt as fuel and is thus lost to agriculture. Not only this tremendous waste, but also the traditional method of preparing and storing the farm yard manure is generally faulty. The cattle dung, together with stable waste and house sweepings, if first collected in the open beak yard and when a cartload has been collected, it is removed to another heap or to an uncovered pit in a common lot outside the village. The loose heaps lie exposed to the sun, with the result that the raw organic matter dried up quickly and does not rot properly Very often, a part of the dry dung is blown off the wind or washed away by rain. Cattle urine is either not conserved or is stored in a defective manner.

However, the loss of nitrogen and mineral elements caused by careless handling can be reduced greatly by using an absorbent bedding for cattle, storing dung in stone or brick -lined pits, mixing large quantities of straw and other vegetable matter with cattle dung, conserved, the loss of soluble mineral elements through seepage is prevented, bacterial decomposition of raw organic matter is encouraged, plant nutrients are made soluble, and nitrogen losses are minimized. Farmers are applying FYM is applied before sowing of a crop.

It must be stressed that the value of farmyard manure in soil improvement is due to its content of principal nutritive elements and its ability to: (i) Improve the soil tilth and aeration, (ii) increase the water holding capacity of the soil and (iii) stimulate the activity of micro-organisms that make the plant food elements in the soil readily available to crops. The supply of organic matter, which is later converted into humus, is a property of farmyard manure. One tonne of cattle dung, can supply only 2.95 kg of nitrogen 1.59 kg of phosphoric acid and 2.95 kg of potash. The use of farmyard manure alone causes an imbalance in nutrition owing to its relatively low content of phosphoric acid. Therefore, to keep the soils well supplied with all the essential elements of plant food in a available from, and also to keep them in good 'heart'. Farmers from Maharashtra are using the bulky organic manures in conjunction with super phosphate and such other artificial fertilizers as contain the particular plant food or foods in which as soil may be deficient or which the crop to be grown may specially require.

Generally farmers from this state are applies 50 t/ha FYM to sugarcane, cotton 20 t/ha ground nut 10 t/ha sunflower 10 t/ha, maize 10 t/ha, bajra 5 t, tomato, brinjal and onion 10t/ha FYM shows a profitable response to vegetable corps, specially potatoes, tomatoes, sweet potato, cauliflower, onion garlic, sugarcane, rice and fruit tress like, bananas, grapes, guava, mangoes etc.

2. Compost manure

Another method of augmenting the supplies of organic is the preparation of compost from farmhouse, and cattle shed wastes of all types. Composting has been advocated and adopted extensively during the past 25 years in Maharashtra in Sholapur, Sangli and Pune District. Composting is the process of reducing vegetable and animal refuse (rural or urban) to a quickly utilizable condition for improving, and maintaining soil fertility. Research conducted in Maharashtra has shown that good organic manure similar in appearance and fertilizing value to cattle manure can be produced from waste materials of various kinds, such as cereal straws, crop stubble, cotton stalks groundnut husk, farm weeds and grasses, leaves, leaf mould, house reuses, wood ashes, litter, urine soaked earth from cattle- sheds and other similar substances. These raw materials are rich in cellulose and other readily decomposable carbohydrates and have a carbon-nitrogen ratio of 40 or more than 1. The direct application of such decomposed, low nitrogen organic matter as manure brings about a temporary deficiency of mineral nutrients (specially nitrates and ammonium compounds) in the soil stimulating the growth of micro -organisms, which in turn, compete with crop plants for available nitrogen, phosphorus and other elements. Hence, before using them as manure, it is necessary to compost or partially decompose them. This process lowers the carbon-nitrogen ratio to about 10 or 12 to 1.

Two methods of composting waste organic materials are usually recommended in Maharashtra. One depends on aerobic and the other on anaerobe decomposition. In both cases, the farm wastes have to be used as bedding for cattle in order to absorb a large portion of the animal urine. In the aerobic process, the used bedding, the sweepings from cattle sheds and some urine - soaked earth form the stable floor are removed every, day mixed with a little cattle dung an two or here handfuls of wood ashes are deposited on a well drained site to gradually build up a low pile, about 30 to 45 cm in height, about site to 5 m in width and of any convenient length. The pile is built up before the start of the rainy season. After the first heavy showers, the wetted material in a 1.2 m strip on each side of the long heap is turned with a rake on to 2.4 wide strips in the middle, so raising the height of the heap to nearly one meter. This process prevents a loss of moisture and ensures a quick start of decomposition. When the heap sinks appreciably and such a sinking takes about three to four weeks, it is given a turning and made into a fresh heap thus mixing the outside material with that form inside. After about a month or more, depending on the incidence of rains the heap is given a final turning on a cloudy or moderately rainy day and rebuilt in the vacant part of the original position, the compost becomes ready for use in about in the rainy season in July.

In the anaerobic process, in Kokan region the mixed farm residues are collected in pits of a convenient size, say 4.5 m x 1.5m x 1m. Each day's collection is spread in a thin layer, sprinkled with a mixture of fresh cow dung (4.50 kg), ashes (140 to 170 g) and water (18 to 22 liters) and compacted. The pits are filled till the raw material stands 38 to 46 cm above its edge, and is then plastered with a 2.5 cm layer of a mixture of mud and cow dung. Under such conditions, decomposing is anaerobic and high temperature to not develop. In soluble nitrogen compounds gradually become soluble and the carbonaceous matter is broken down into carbon dioxide and water. The loss of ammonia is negligible, because in high concentration of carbon dioxide, ammonium carbonate is stable. The plastered pit also prevents the fly nuisance. The compacted moist material becomes composted in about four month without any further attention. The well made compost contains 0.8 to I per cent of nitrogen and has all the good properties of farmyard manure. It can be used in the same way as the latter. The anaerobic process is particularly used by gardeners or near cities and towns in Maharashtra. The com posted manure is applied before sowing of crop.

3. Town compost

In recent years, large scale composting of town refuse and night soil in properly constructed trenches away from human habitations has been taken up successfully, by the municipalities of Pune and Mumbai and many large and small towns in Maharashtra. Trenches, I to 1.2 m wide 75 cm deep and of convenient length, are filled with successive layers of night soil town refuse and earth, in this order. The compost gets ready in about three months. Soil prejudice against the use of this valuable compost has disappeared and town composting is almost is being rapidly adopted in other localities all over India With a suitable codification, such as the provision of trench latrines, it can be taken up in village. too.

(i) Preparation of synthetic compost

Such types of compost are prepared by some firm at Pune and Mumbai. the basic principles underlying the microbiological process of conversion of organic wastes into manure have been extended for preparation of synthetic composts, using inorganic fertilizes as sources of nitrogen for decomposition of carbonaceous   materials. It has been found that organic nitrogen in the form of dung required by microorganisms can completely substituted by inorganic nitrogen compounds like ammonium sulphate and urea, which are utilized equally effectively for decomposition and well fermented manure, resembling farmyard manure, can thus can be obtained? This opens up a vast field for utilization of large quantities of various organic waste materials where supplied of dung are either short of requirements or not available at all as on mechanized farm. The basic principles of C/N ration in manure preparation can be applied to add nitrogenous fertilizers in sufficient quantity to reduce the ratio to about 30 : 40 : I and then allow the material to be decomposed and disintegrated until the C/N ratio is reduced to 20: 1. The adco process of preparation of synthetic compost worked out by Hutchinson and Richards is based on this principle.

This process of production of compost from organic wastes of wide C;N ration is worked with fertilizers like ammonium sulphate or urea. They provide complete or partial nitrogen for microbial decomposition in place of natural materials like dung or night soil which may not be available at all or may be in short supply. Nitrogen is added to make up a total of 1.2 per cent on dry weight basis of the material. Thus straw containing 0.4 per cent nitrogen requires additional 0.8 per cent nitrogen, i.e. about 4 kg ammonium sulphate per 100 kg of straw. For neutralizing acidity, which may develop in this system, the process includes addition of lime at the rate of 5 kg to the above material; the process includes addition of lime at the rate of 5 kg to the above material. The material to be composted is spread out in layer in a heap or pit in sufficiently moistened this is sprinkled with the fertilizers solution and then with lime. Superphosphate may to added to fortify the phosphatic contents of the manure., The treatment is continued layer wise until the heap pit is filled to size and allowed to ferment. The manure becomes ready for application in about four to six months and resembles farmyard manure in this application soil and plant growth. This manure is applied before sowing of a crop.

(ii) Mechanical Compost Plants

Although the major quantities of compost at present used in agriculture are prepared in villages in Maharashtra either by traditional or improved methods (Indore or Bangalore methods), these methods, cannot be suitably used for processing of large quantities of organic refuse of bigger cities. The mechanized/ semi mechanized plants will serve tow pertinent objectives -sanitary disposal of city refuse of production of organic manure. Moreover, Indian city wastes contain about 50 to 80 per cent of compostable organic materials which can be processed conveniently in compost plants. The intake capacity of these plants already setup in about 125 to 200 tonnes garbage per day and can produce 60 to 70 tonnes of finished composts per day if the plant runs in shifts. The costs of production is rather high due to mechanism of the processed and at some places in Maharashtra the plants were running at underutilized capacity due to non availability of required quantities of garbage.

The compost prepared form city wastes are bulky and low in plant nutrients. It contains 0.5 to 0.6 per cent N, 0.6 per cent P_2O_5 and 0.5 per cent K_2O with a C/N ratio of 20: 1. There is a need to enrich this manure in respect of nitrogen and phosphorus and reduce its bulkiness involving low cost technology. The application of 10 kg low grade rock phosphate containing 20 per cent P_2O_5 per tonne of organic wastes (crop wastes and leaf fall, etc.) and in inoculation with microbial fertilizers improved the nitrogen, phosphorus and humus content appreciably this can be adopted in improving the compost prepared by mechanical compost plants.

4. Green Manure

In Maharashtra green manuring is followed in same districts. It consists in the growing of a quick growing crop and ploughing it under to the incorporate it into the soil. The green manure crop supplies organic matter as well as additional nitrogen, particularly of it is a legume crop which has the ability to acquire nitrogen from the air with help of its root nodule bacteria. A leguminous crop producing 8 to 25 tonnes of green matter per hectare will add about 60 to 90 kg of nitrogen when ploughed under. This amount would equal an application of three to ten tennes of farmyard manure on the basis of organic matter and its nitrogen contribution. The green manure crops also exercise a protective action against erosion and leaching. The crops most commonly used for green manuring in Maharashtra State are Sunnhemp *(Crotalariajuncea)*, dhaincha *(Sesbania aculeata)* cluster bean *(Cyamposis tetragonoloha)*, cowpea *(Vigna catjang, V sinensis)*, horse-gram *(Dolichos hi/lorus)*, berseem or Egyptian clover *(Trigolium alexandrinum)*. Lentil *(Lens esculanta)* is used for green manuring to paddy. Sown in late autumn, it is said to provide a winter cover and make new growth in early spring for ploughing under before the sowing of paddy. Sunnhemp is the most outstanding green manure crop. It is well suited to almost all parts of the Maharashtra and fits in well with sugarcane, potatoes, and garden crops. Dhaincha does well on alkaline and water logged soils. Berseem is well suited or orchards and the irrigated crops of cotton and sugarcane sown in spring or early summer. The best legume to be used as a green manure in any given locality is manually the one most suited to its soil and climatic conditions.

Very often, berseem and Lucerne *(Medieago sativa)* and sometimes sunnhemp are grown partly for fodder and partly for green - manuring in Maharashtra. In the case of annual crops of berseem, one or more cutting is taken for use as green fodder. Lucerne which is allowed to grow for two to three years is cut seven to eight times the same purpose. In the case of sunnhemp, the tops are fed to cattle. In all these instances, the residues (roots and stumps) are incorporated into the soil. These crop resides contain considerable amounts of nitrogen, phosphorus, potassium and other mineral nutrients, besides organic matter, In the case of orchards, the annual green manure-cum-forage crop should be grown at such a time as to interfere the least with tree growth and fruit development.

In recent years, extensive efforts have been made in Maharashtra to plant *Clvrieidia maculata* and *Sesbwlia speeioea.* On the borders of paddy field or in other vacant spaces to provide green leaf for manuring the paddy crop in Konkan region of Maharashtra Grown from seedlings or rooted stumps, the paddy crop planted 2 m apart, each Glyricidia plant is said to give annually low cuttings each of about 6 to 12 kg of green leaf. It does well-in both red and black soils. Similarly, Sesbania species seedli!1gs planted 10 cm apart on paddy borders produce 1000 to 2500 kg of green leaf for manuring 0.4 ha of paddy. Only 115 g of seed is needed to provide seedlings sufficient for border planting of two hectares of paddy. In certain other paddy growing area, *Pongamia pinnata* (Karanj) and other trees yielding large quantities of leaves are planted for use as a green manure. For the proper rotting of the green manure, it is necessary that the green material should be succulent and there should be adequate moisture in the soil. Plants at the flowering stage contain the greatest bulk of succulent organic matter with a low *carboni* nitrogen ratio. The incorporation of the green manure crop into the soil at this stage allows a quick liberation of nitrogen in the available form. With advancing age, the percentage of carbonaceous matter in the plants increases and that of nitrogen decreases. If the material with a wide carbon--nitrogen ratio is ploughed under micro-organisms bring about its decomposition, draw upon the released nitrogen and about its decomposition, draw upon the released nitrogen and about mineral nutrients and cause the temporary nutrient deficiency. This green manure is buried in a soil and after complete decomposition crop is sown.

The increase of yield after green manuring is usually of the order of 30 to 50 per cent. The fertilizing value of the legume crop can be increased a great deal by maundering it with superphosphate. This practice not only increases the phosphorus content of the green manure plants, but also encourages plant growth on the whole, thus converting an inorganic fertilizer into organic manure. Green manures have a marked residual effect also. Glricidia contains 8.5 per cent carbon and 0.49 per cent nitrogen. The yield of leafy material per plant according to the observations was 22.5 kg after the tenth year. The method of application of green manure to sugarcane was in situ in all the cases. The most commonly used green manure in Maharashtra is sunnhemp. When sunnhemp is grown in kharif season it will gives 1.52 t/ha green matter and it will add 84 kg N/ha. When Dhaincha is grown in a kharif season, it will give 1.44 t/ha green matter and it will add 77. 1 kg/ha. When cowpea is grown in a kharif season 1.08 t/ha green matter and it will add 56.3 kg N/ha.

5. Poultry manure

The poultry manure is used by the farmers from a Sholapur, Satara, Sangli and Ahmednagar districts of Maharashtra. This is rich organic manure, since liquid and soil excreta are excreted together resulting in 'no urine loss. Poultry manure ferments very quickly. If left exposed, it may lose up to 50 per cent of its nitrogen within 30 days. Poultry manure can be applied to the soil directly as soon as possible. After application, it should be worked into the surface of the soil. If the droppings come from the cages or droppings pits, superphosphate may be added to these at the rate of 1 kg per day, per hundred birds. This improves the fertilizing quality and helps the control of flies and odour.

In Konkan region of Maharashtra, groundnut has been introduced during the last decade only and the area covered under the crop has increased to 5800 ha. The fanners of the region cannot afford to apply recommended dose of fertilizers to the crop i.e. 25 kg N+ 50 kg P_1O_5 ha-^{I}. It is, therefore, necessary to supplement the chemical fertilizers with poultry manure which is available in plenty in the some districts of Maharashtra. The average chemical composition of the fresh poultry manure is 75% moisture, nitrogen 1.7 %, phosphorus 1.15 & and potassium 0.48%. This manure is applied before sowing of a crop.

Poultry manure contains essential plant nutrients such as nitrogen, phosphorus, potassium, calcium magnesium, sulphur, boron, zinc, copper, manganese, iron etc. which are necessary for increasing the yield and quality of groundnut. Since solid and liquid portion the poultry excreta are excreted together, poultry manure is a concentrated source of nitrogen and phosphorus. If poultry manure is added in combination with chemical fertilizers, it will supplement all these nutrients to groundnut, thereby boosting up not only the production of ground nut, but improving its quality also. Alike the yield of pod and haulm, the uptake of N,P,K, Zn, Cu, Mn and Fe increased significantly with increasing levels of fertilizers and poultry manure. Addition of poultry manure. Addition of poultry manure from 1 to 3 t/ha with zero, half and full doses of fertilizers resulted in successive increase in the uptake of macro as well as micronutrients. Highest uptake of said nutrients was noticed with recommended dose.

Different techniques for making composting

Modern agriculture depends upon the external application of plant nutrients to meet crop needs. Soil reserves cannot provide large amounts of nutrients needed year after year to harvest the quantum of crop produce required for increasing human population. It is now clear that no single source of plant nutrients can meet the total nutrient demand and sustaining crop productivity depends upon judicious use of organic as well as mineral fertilizers. Most cultivated soils in tropical climates arc poor in organic matter due to high rate of decomposition and repeated cultivation. Recycling of organic wastes in agriculture brings in the much needed organic matter to the soils. Since most recyclable wastes arc 'Organic". They directly add organic matter and the plant nutrients contained in them while nutrient input improves soil fertility; the organic inputs has a profound and vital role to play in improving soil physical properties. Thus organic waste recycling leads to an improvement in the overall soil productivity; of which soil fertility is a key component.

Many factors are responsible to increase the yield of crops. Amongst them essential are to provide food nutrients to growing crops. This need can be fulfilled through chemical fertilizers and organic manures. Nutrient supplementation through chemical fertilizers and its availability to farmer at a reasonable cost is a real problem. The option remains in use of quality compost. Hence. it is necessary to prepare compost with a specific technique using agricultural wastes. A lot of work have been done by different workers on composting by using different methods and proved that it has beneficial effect on plant growth, yield and properties of soil. Addition of organic manures may increase the population of beneficial microorganisms and enhance their activities. so that nutrients are made available to crops. The organic wastes available in different forms can be used for composting. In our country, nearly 1000 M tones of agricultural waste, rather than other sources is available, even through productivity of compost is a constrain and availability at high prices. If it is managed properly the productivity and prices can be controlled. A time has come to think seriously about the use of compost to maintain soil fertility. To obtain good quality compost following points should be considered on top priority.

1. C:N ratio

Composting takes care of most of the pathogens present in the substrates. Composting involves mineralization of waste biomass components by microorganisms. The number and variety of microorganisms present during composting is very large and their activity is intense due to the variety of substrates present in the waste. The decomposition of compostable material depends upon its chemical composition and factors like C : N ratio, particle size, pH, minerals, temperature, moisture and aeration etc. If C : N ratio of composting material is 30: 1 or less than this, decomposition process will be faster, if it is more than 30 : 1 process is slow or delayed for completion. Hence to avoid such circumstances addition of 2 kg Nitrogen through Ammonium sulphate or urea and 2-3 kg of super phosphate helps for 1000 kg compost preparation. C: N ratio of 20 : 1 or less of the final product is considered optimum, to avoid nitrogen immobilization which affects plant growth.

2. Moisture content

Decomposition process is totally depend upon moisture content of substrates; initially 60 to 70 per cent moisture is required in pit, every 15-20 days after filling watering is needed. When compost is completely matured, it must hold 30 - 40 per cent moisture. Excess moisture is avoided in pit which forms fermentations and resulted in lowering specific temperature and decreases decomposition process which leads to the formation of low quality compost.

3. Temperature

Most of the microbes that function at temperature between 10°C and 45°C, to initiate composting process and as temperature increases as a result of oxidation of carbon compounds; thermopiles (micro organisms that functions at temperature between 45 °C and 70 °C) take over. Temperature in compost pit typically follows a pattern of rapid increase to 49 °C to 60 °C within 24 to 72 hours of pile formation and is maintained for several weeks. This is a active phase of composting; in which early degradable compounds, oxygen are consumed, unwanted pathogens and weed seeds are killed and phytotoxins (organic compounds toxic to plants) are eliminated. During thermophilic, active composting phase, oxygen must be provided throughout the compost pile. By adopting new technique, methods like, NADEP, chimney (JNKY Jabalpur method) and recently developed, named as MAU-method (Dept. of ACSS -MAU, Parbhani) maximum aeration is provided through perforated PYC pipes laid in pit.

4. Micro-organisms

Composting is a dynamic process involving rapid succession of mixed microbial population. The precise nature of succession and number of microorganisms at each stage depends on the composting material and upon the proceeding organisms in the succession. A range of bacteria fungi and actinomycetes have been recorded in composting. The relative preponderance of one species over another depend up on the constantly changing available nutrient supply, temperature and substrate conditions, mesofauna such as mites, sow bugs, worms, spring nails, ants, nematodes and beetles do most of the initial mechanical breakdown of organic materials in smaller particles. Mesophilic bacteria, fungi, actinomycetes and protozoa initiate decomposition process, as a result of oxidation of carbon compounds. Also degradation process is carried out by different kinds of heterophytic microorganisms which derive their energy and carbon requirements from the decomposition of carbonaceous materials. For every 10 parts of carbon, 1 part nitrogen is required for building up their cell protoplasm. Under anaerobic conditions, microorganisms breakdown organic materials by a process of reduction in the absence of 02 first, special group or acid producing bacteria and facultative heterotrophs degrade organic matter in to fatty acids i.e. aldehydes and alcohol; then a group of bacteria convert the intermediate product to methane.

5. Use of Microbial Inoculants

During development of composting systems, research also has been done on the use of inoculants to hasten the process of composting. Inoculation with *Trichoderma viridi* helps to decompose cellulose, lignin of substrate to a fast rate. Inoculation with *Trichruspiralis* and *Paecilarnycej lusisporus* help to produce compost with lower C:N ratio from sugarcane trash and rice straw. The inoculation with nitrogen fixing bacterium *Azatobacter*, phosphate solubilizing fungus *Aspergillus awamori* and *cellulytic* fungi coupled with the addition of rock phosphate helps to produce compost residues, plant residues in a shorter period and with high N and P content.

6. pH

During decomposition process of composting, organic acids are released and pH of compost pile is decreased and due to which the rate of decomposition is lowered down. To minimize the effect of such organic acids the pH is maintained to 7.00 to 7.5. For this 5 per cent lime or ash is to be added in compost pit.

7. Final product (Matured compost)

Minimum and maximum 3 to 4 months period is required to complete the process of composting. At the end, final product attains only 50 per cent weight of the added fresh material and it results in 15 : 1 to 20 : 1 C : N ratio. It appears dark brown in colour, soft textured, odourless, which is very easy to spread over field. All above said properties of compost indicate a complete and good quality product. A partial decomposition may lead to the formation of low quality compost which shows C:N ratio above 20: 1. Application of such compost fixes nitrogen from soil and the availability of soil nitrogen is delayed for some time to growing crops.

Preparation of pit

In general, pit should be 3 M (L) x 2 M (W) x I M (H) in size; length of pit can be extended as per the availability of composting material with a maximum limits of 6 M to 10M. Procurement floor of the pit may be compacted by any mean, so as to avoid water and nutrient losses from compost pit during decomposition process.

Materials used for composting

1. Agricultural wastes: Wide variety of crop residues, animal wastes, *kharif* weed Parthenium, grasses (before flowering stage), glyricidia; soybean trash, wheat straw, sugarcane trash, maize stubbles, all agricultural wastes can be used as a substrate for composting.

2. Cow dung slurry : (Act as self inoculums)

3. *Trichoderma viridi* (Inoculants)

4. Fine soil (starter material)

5. Urea

6. Single Super Phosphate

7. Five per cent lime and ash is recommended for 1000 kg substrate and 3 m x 2 m x I m (L x B x H) Size pit.

Adoption of New technique for filling of substrates in pit

Recently new technique is preferred to fill substrates in a pit so as to get quality compost within short period of time and high nutrient content. Specifically following steps are recommended.

1. Substrate filling technique

(i) Fifty kgs of different agricultural wastes or other than this to 4 to 6 cm in size (small size particles have more surface area, which provides ample of aeration and free movement of water in compost pile) should be spread uniformly in pit to the extent of 15 cm thickness of each layer.

(ii) Add 20 liters of water per layer if material is dry in nature.

(iii) Spread 20 kg of fine soil per layer which act as starter material for composting and biblical references to farmer composting.

(iv) Add 20 kg of fresh cow dung slurry made from 40 liters of water; sprinkled on a layer equally.

(v) Add 3 kg of nitrogen through (Ammonium sulphate or urea) and Single Super Phosphate 3.5 to 6 kg per pit. Addition of SSP helps for saving 20 to 40 per cent nitrogen Loss as $NH/$ during decomposition process and to enrich the compost for "P" content.

(vi) Add 50 g of *Trichoderma viridi* directly on substrate per layer or it can be added through culture, made on I kg of wheat straw (2 to 3 days before composting) which decomposes fast cellulose and lignin of substrate and time required for preparation of compost is reduced to the extent of 20 to 25 days as compared to other technique.

(*vii*) Such ten layers are to be prepared, 6 layers inside the pit and 4 layers above the pit. At the end of layer; it should turned to dome shape and plastered with fine soil and cow dung slurry. 60 to 70 per cent moisture is maintained in compost pile, addition of water every 15 to 20 days is followed. The watering should be stopped one month prior to maturity stage of composting.

In general by adopting new composting technique we can obtain quality compost within 90-100 days with a high nutrient content in N, P, K and narrow C : N ratio i.e. 20 : 1 or less than this.

Innovative Composting Technology for Agricultural Wastes

The importance of organic manure in agriculture is known since ancient times. The organic manure is the life of soil and if neglected the fertility of soil would not be maintained. The occurrence of multi nutrient deficiencies and overall decline in the productive capacity of soil under intensive fertilizer use has been widely reported.

At present, most optimistic estimates show that about 2S to 30. per cent nutrient needs of Indian agriculture can be met by utilizing various organic sources; so as to overcome these problems, it is necessary to focus on production of quality compost, based on new technology. In India, since 1930 different methods of composting were followed. Amongst them Indore and Bangalore methods are well known and widely adopted. Procurement of compost through these methods was observed very poor in quality, nutrient content and more time requirement for completion of compost. These facts clearly indicate lack of technical knowledge of composting. By adopting new technology a known Gandhian Shri. Namdeorao Pandhar Pande from Pusad, Dist. Yeotmal, (MS) have developed NADEP method of composting which was found superior over Indore and Banglore methods. In 1999, Prof. Sharma and Rawat from JNKV, Jabalpur (MP) worked on the line of composting and developed new method, named as Chimney method and proved to be superior as compared to NADEP method. The adaptability for these methods is uneconomical to Medium and Low holding farmers due to high cost of construction and maintenance requirement for every term of composting. In view of the above facts, it was therefore, thought worthwhile to think on new technique of composting. As composting in pit method requires more aeration for microbial decomposition process. For this achievement Prof. N. R. Khan from MAD, Parbhani (MS) has developed composting method, based on new technique named as MAD Method. Research attempts have been made under the guidance of Dr. S.D. More, Head, Department of Agricultural Chemistry and Soil Science, MAD, Parbhani. To facilitate more aeration and equal distribution of water, in pit; use of perforated PVC pipe is done in MAD, Method. This new technique has established the production of quality compost with high nutrient content and less duration is required to the extent of 20 to 2S days for completion of composting process.

Energy Source of Microbes in Recycling of Organic Wastes

The supply of carbon relative to nitrogen (C:N ratio) determines whether net mineralization and immobilization of nitrogen will occur. Mineralization is conversion of organic nitrogen to mineral forms (i.e. ammonium and 'nitrate): immobilization is incorporation of mineral into microbial biomass as a general rule, if the CN ration is greater than 20: 1, microbes will immobilize nitrogen into their biomass. If C: N is less than 20: 1, nitrogen can be lost to the atmosphere as ammonia gas, causing odor. In general, green materials have lower C: N ratios than woody materials or dead leaves do, and animal wastes are more nitrogen rich than plant wastes are. The complexity of the carbon compounds also affects the rate at which organic wastes are broken down. The ease with which compounds degrade generally follows the order carbohydrates > hemicelluloses > cellulose + chitin > lignin. Fruit and vegetable wastes are easily degraded because they contain mostly sugars and starches. In contrast, stems, nutshells, bark, and tree limbs and branches decompose more slowly because they' contain cellulose, hemicelluloses and lignin.

Moisture Requirements for better composting

Low moisture content impedes the composting process, because microbes need water. Low moisture also makes compost piles more susceptible to spontaneous combustion, because moisture content regulates temperature. Moisture content in excess of 60% means pore spaces in the compost pile are filled with water rather than air (oxygen), leading to anaerobic conditions. Feedstock with different moisture holding capacities can be blended to achieve ideal moisture content. Carbonaceous materials such as newspaper and wood by-products such as sawdust are often used as bulking (drying) agents.

Important environmental considerations composting Organic wastes

Minimum oxygen content of 5% should be maintained for aerobic composting. As microbial activity increases in the compost pile, more oxygen is consumed. If the oxygen supply is not replenished, composting can shift to anaerobic decomposition, which often results in bad odor. Bacterial decomposers prefer pH in the range of 6.0 to 7.5 and fungal decomposers prefer pH of 5.5 to 8.0. Certain materials, such as paper processed food wastes can lower pH. If compost pH exceeds 7.5 gaseous loss of ammonia is more likely to be occured. The particle size of organic wastes for composting is important for microbial activity and airflow in the compost pile. Smaller particles have more surface area per unit volume; therefore, microbes have greater access to their substrate. Thus, grinding of feedstock before composting can accelerate the composting process. However, resulting in anaerobic conditions.

Process of composting

Mesofauna such as mites, sowbugs, worms, spring nails, ants, nematodes, and beetles do most of the initial mechanical breakdown of organic materials into smaller particles. Mesophilic bacteria, fungi, actiomycetes, and protozoa (microbes that function at temperature between 10°C and 45 °C(50 °F-113 °F) initiate the composting process and as temperature increases as result of oxidation of carbon compounds, thermopiles (microorganisms that function at temperature between 45 °C and 70 °C (°F - 157 °F) take over. Temperature in a compost pile typically follows a pattern of rapid increase to 49 °C to 60 °C (120 °F -140 °F) within 24 to 72 hours of pile formation and is maintained for several weeks. This is the active phase of composting, in which easily degradable compounds, oxygen is consumed, pathogens (e.g., *Escherichia coli, Staphlococcus aureaus, Bacillus subtilus, Clostridum hotulinum*) and weed seeds are killed, and phytotozins (organic compounds .toxic to plants) are eliminated. During the thermophilic, active composting phase, oxygen must be replenished by mixing, forced aeration, or turning to the compost pile.

Composting of Garden and Kitchen waste

Landfill banning of municipal organic wastes such as leaves and grass clippings in the late 1980s, along with increased homeowner interest in recycling and organic gardening, has been a boon for home composting. Home composting is one of the most cost-effective organic material management strategies because, it eliminates the costs of collection and processing.

Organic waste suitable for home composting includes grass clippings, hay, straw, sawdust, wood chips, kitchen waste (e.g., fruit and vegetable peels and rinds, tea bags, coffee grounds, egg shells), leaves, and animal manure (e.g. foultry, cow, horse). It is best to combine dry, high-carbon materials (e.g., woody materials, straw, hay) with wet, high-nitrogen materials (e.g., grass clippings, food scraps, manure) to optimize the C: N ratio, moisture content, particle size, and pile porosity.

Some Important Methods of Composting

Indore Method of Composting of Agricultural Wastes

The waste materials such as plant residues, animal wastes, vegetable wastes and weeds can be composted with the Indore method. First the waste materials are chopped into small pieces of 5-10 cm size and are dried to 40-50 per cent moisture before stacking. Then, they are spread in layers of 10-15 cm thickness either in pits or in heaps of one m width, 4-6 m length and 1 m depth. The heap is properly moistened with dung, using earth or night soil. Sufficient quantity of water is sprinkled over the heap to wet the composting materials to the level of 50 per cent moisture. Periodical turnings, usually three times at 15,30 and 60 days after filling, are given to aerate and the material is covered with a thin layer of soil of about 2-3 cm thickness. Under the aerobic process of decomposition, losses of organic matter and nitrogen are heavy (40-50 per cent at initial stage). This process, however, involves considerable labour in the preparation of the heap and periodical turnings and so becomes expensive and impracticable when large quantities of materials are to be processed. The site selected must be at a high level to prevent rainwater stagnation. It should be nearer to cattle shed and water source for easy transporting. The compost obtained by this Indore method would have a composition 0.8 per cent N, 0.3 per cent P and 1.5 per cent K_2O.

Activated Compost

Fowler developed this technique in which fresh materials are incorporated in an already fermenting heap so that quicker decomposition can be obtained with already established microbial population. This method is useful, particularly when offensive materials like night soil are to be quickly and effectively disposed off.

Banglore Method

Pioneering work in preparation of manure in pits was carried out by Acharya (1939), particularly on the utilization of town residues and night soil. This process is otherwise called as hot fermentation method of manure production. In this method, the compost production depot is located on the city outskirts to transport town refuse and night soil to the pits. The depots normally accommodate about 200 trenches with a spacing of 1 – 5 m between trenches. First the refuse is to make 15-cm height layer. Then night soil is discharged over this and spread to a layer of 5cm. After filling the pit to refuse and night soil in alternate layers, the pit is filed to 15cm above ground level with a final layer of refuse of 15 cm on the top. This may be dome shaped and covered with the thin layer of soil with red earth or mud to prevent moisture loss and breeding of flies. Sludge water, if collected carts as in some towns, may be emptied over the layer of refuse. This system provides a method of disposal of any kind of waste, including slaughterhouse waste, carcasses of animal, sewage, etc.

The materials are allowed to remain as such without any curing and pot watering for about three months. The decomposition of dumped materials in pits takes place largely in the absence of sufficient air except in the surface layer. Thought the decomposition is comparatively slow, high temperature is not developed in the lower layers. Since the material does not receive any turnings, decomposition into a homogenous mass of manure does not take place, even then, the C/N ratio is reduced to less than 20: 1 in about six months and the manure is ready for use. As there is no watering and turning, it is suitable to areas having low water availability and with scarce labour. When pits or trenches are not available for composting, town refuse and night soils can be composted in above ground heaps of 1 m height, 1 m width and of any convenient length. By placing refuse and night soil in alternate layers as in trenches and adding the final refuse on the top (Gaur *et al.*, 1984). In this method, the material decomposes more quickly than in pits and can be used after 3-4 months. The compost obtained by this method would contain 1.5 per cent N, 1.0 per cent P and 1.5 per cent K_2O.

Nadep Compost

A Survodai Leader Namdeo Pandhripande of Yeotmal district of Maharashtra proposed this method. In this method, plant wastes, dung slurry and clay soil is used as raw materials for composting. The process is similar to heap method of composting, but is done in brick lined enclosures provided with air holes on all sides. However, this method has the disadvantage of using large proportion of soil. It is observed that compost obtained by this method had increased available nutrients.

Coimbatore Method

It is anaerobic degradation followed by aerobic process. First, pits of 4m length, 2m width and 1 m depth is formed in which crop residues or farm wastes are filled to a thickness of about 15cm. Over this layer, cow dung slurry to enhance the rate of biodegradation is applied to a thickness of 5cm. above this layer, 1 kg of bone meal, or rock phosphate to minimize the nitrogen loss and to add phosphorus, is applied. Thus, application of crop residue/farm waste, cow dung slurry, bone meal and rock phosphate in alternate layers is repeated till the height reaches 0.5 m above the ground level. The above ground portion is covered with red earth or mud to prevent the rain water entry and it becomes an anaerobic process. After 30-35 days, the material is turned and it becomes an aerobic process. The compost will be ready within five months.

Synthetic compost

The principle behind this procedure is that addition of dung to supply organic nitrogen required by microorganisms, can be completely substituted with fertilizer N compounds like urea and ammonium sulphate which are utilized effectively, for decomposition of carbonaceous compounds. The basic principle of

C/N ratio in manure preparation can be applied in synthetic composting as well be adding nitrogenous fertilizers insufficient quantity to reduce the ratio to about 30-40 per cent and then allowed to decompose. This method is highly suitable to the areas where supplies of dung are either short of requirements or not available at all and large quantity of organic waste materials is available in the farm.

Either pit or heap method can be used for composting where the organic waste material is spread out in thin layer and sufficiently moistened. Super phosphate may be added to fortify the phosphorus content of the manure. Sprinkling of fertilizer solution followed by lime application may be done. This process is continued layer-wise until the heap or pit is filled and then, the heap or pit is allowed to decompose for a period of 4-6 months by which time; it is ready for field use. It resembles FYM in its effect on soil and plant growth.

Windrow composting (Leaf compost)

Windrows are prepared as they allow efficient material handling, provide good aeration and allow sufficient absorption of water. First, windrows of 2.5-4 m width at the base, 2.5-3 m height and of any convenient length based on the availability of leafy materials are formed. It is better to use higher windrows for better- decomposition of leaves at the base resulting in aerobic conditions. When the moisture content of incoming leaves is low, it is desirable to add sufficient quantity of water to maintain 40-60 per cent moisture. If the C/ N ratio of the leaves is high (30), amending with sewage sludge, urea or grass clippings may be done. Periodical mixing may be given for good aeration. The leaf compost with a neutral pH (6-7) will be ready within 6-9 months under optimum environmental conditions. The use of finished compost (leaf mold) as covering material (10-15cm) in the subsequent perorations of leaf compost is good to supply a heavy inoculate of microorganisms as in the case of activated compost' (Eberhardt and Pipes, 1972).

Enriched Compost by Acceleration

For getting better compost, it is necessary to add nutrients, N, P and K which hastens the process of composting. Addition of *mesophilic celluloytic* fungi, or *Coprinus ephemerus,* a cellulose decomposer to compost pit may reduce the composting period by one month with an improvement in compost quality. Inoculation with *Asperglilus niger or Pencillium* sp. Increase the total N content, available P and humus content of lower stalk plus wheat straw compost and Jamin leaf compost (Gaur et al., 1982). Further, this may also prove beneficial in the case of *in situ* application of plant residues where thermic rise is unlikely to be more than 40-45 ^{0}C. Similarly, inoculating the compost previously amended with rock phosphate with cultures of *Azotobacter chrococcum* and the phosphate solubilizing strain *Asperigillus avalllori* increases the total nitrogen and humus content.

Vermi - composting

In Asia, vermicomposting is being effectively practiced in Philippines. The earthworms, *Lumbricus rubillus* and/ or *Perionyx excavator,* are used for composting. Organic matter can be converted by earthworms into 300 t of compost, rich in N,P, K, ea and Mg contents with increased bacterial and actinomycete population. This method of composting can be suggested for sludge management.

Animal waste composting

Animal manures of different kinds are available in huge quantities in India. If they are not used properly, it will lead to water pollution through rains and runoff water, air pollution in the forms of disagreeable odours and human health hazard through multiplication of flies and other insects. Animal waste composting is an effective and environmentally safe method to eliminate odour, to control flies and other insects, to reduce mass and volume, and to handle the material easily with agreeable smell. The type of animal, quantity and quality of feed, housing and waste handling techniques will decide the characteristics of the manure. In general, beef and poultry manures are rich in nitrogen and undergoes decomposition more effectively with lower moisture level content. Addition of carbonaceous material such as straw, wood bark, saw dust, or corn cobs can often help the composting characteristics of manure. They can be used to reduce moisture content and to raise the C/N ratio, as it is usually practiced in all villages. However, under Indian condition, conjoint composting of the manure slurries with plant residues is more viable and profitable. Mustuzaki (1977) produced animal waste compost with 30-40 per cent moisture content in about 2 weeks, by regulating the moisture content of raw materials to 55-65 per cent. Using this low moisture compost (seed compost) as moisture regulator and inoculums of useful microorganisms, he devised method of continuous composting of fresh animal wastes in which high quality compost can be obtained every two week, regardless of weather conditions and with no use of supplementary energy.

Oil Palm Waste Composting

The palm oil industry plays a major role in the economic development in South East Asia, and is slowly gaining momentum in Southern India. Compost can be prepared by using equal proportion of oil palm waste and wheat straw mixed with small quantity of deep litter chicken manure. Before composting, the waste is analyzed for its physical and chemical characteristics.

The ingredients mentioned above are thoroughly mixed and wetted to give compost mixed with no draining liquid. The prepared compost mixture (800g) is then placed on a perforated stainless steel platform within each flask and the flasks immersed in the water bath such that the water level is above the enclosed substances. The thermostat is initially set at 45-50 ^{0}C, to allow a gradual buildup of the naturally occurring microflora and each flask is independently aerated with humified air at 4 1/hr throughout the composting period. When the compost temperature, has attained 50 ^{0}C to pasteurize the material within the flasks. After 6 hr's at this temperature, the compost temperature is raised to 53-55 ^{0}C where is held for further 2-3 days until the concentrations of ammonia fall below 20 PPM which is determined using Dragger gas detector. The purpose of adding wheat straw is to improve the water holding capability and chicken manure to increase the total N content of the compost. Compost mixture of 500 g wheat straw + 500 g oil palm waste + 250 g chicken manure + 30 g gypsum will have total N 1.1 per cent 36.6 per cent total C and a C: N ratio of 33.3: I. It can be supplemented with carbohydrate sources such as sugarcane baggase or molasses to activate the resident microflora.

Phospho-compost

Phospho-compost can be prepared by mixing farm wastes/crop residues, cattle dung, soil, chopped grasses and tree leaves with Mussoorie rock phosphate at the rate of 30 per cent of the computable material. This mixture is made into slurry so as to provide adequate moisture and after uniform mixing; the slurry is allowed to decompose in a compost pit for about 60 to 90 days. (Mishra et al., 1984). Moisture is maintained at 70 per cent throughout the period of composting and the compost will be ready for use in 60-90 days. Normally, this phosphocompost is prepared by mixing organic wastes and Mussourie rock phosphate in the ratio of 7 to 7.5 : 2.5 to 3.0.

Enriched compost using sugarcane trash

In India, 220 million tons of canes are produced annually. At the rate of 10-20% of cane harvested, about 22-44 million tons of cane trash would be produced annually. In addition to this, every 100 tonnes of cane crushed in factory leaves out 30 tonnes of baggase and 3-4 tonnes of pressmud. All this if scientifically utilized to prepare compost and apply for fields one can greatly reduce the cost on inorganics and mange the trash well. Today most of sugarcane cultiators burn the trash in the field itself.

Disadvantages of trash burning

1. Smoke pollutes the air and concentration of CO_2 in atmosphere increases greatly

2. Burning reduces the great volume of dry organic waste into little ash and along with it nutrients also burnt / lost.

3. Burning of trash lying on field surface increases the soil temperature greatly thereby kills crop and soil beneficial microbes and

4. Eye buds of ratoon crop stumps present in soil are also burnt thereby germination of ratoon crop is reduced.

Hence, by recycling the cane trash as manure, the productivity of the sugarcane crop can be enhanced. On an average every tonnes of dry trash is known to contain N, P and K equivalent to 8-10 kg of urea and SSP and 11-12 kg of MOP. Kumaraswamy (2000) there are several methods available to better use the sugarcane trash and prepare enriched/nutrition compost.

Reinforced compost from sugarcane trash and pressmud

First, sugarcane trash is spread to a thickness of 15 cm over an area of 5-7 m x 3 m depending on the area available. Over this, pressmud is spread to a thickness of 5 cm. The fertilizer mixture containing Mussourie rock phosphate, gypsum, and urea in the ratio of 5:4 : I is sprinkled over these layers at the rate of 10 kg/100 kg or trash. Enough water is sprinkled to keep it moist. This has to be repeated till the height of heap rises to a height of 1.0 -1.5 m above the ground level according to the convenience. Cow dung slurry may be used instead of water to moisten the layer. This is covered with a layer of soil and pressmud at 1: 1 ratio to thickness of 15cm or with cow dung paste. Water is sprinkled once is 15 days. A thorough turning is given after 3 months and again heaped. Another turning is given after a month. The trash is composted and is ready in five months. This method increased the material value of trash compost by increasing N, P and Ca contents. It also brings down the C:N ratio by 10 times as compared.

Pit method of enriched composting

Depending upon the convenience of the farmer, a pit of about 10m long, 5 m wide and one meter deep is dug and about a half tonnes of dry trash is spread at the bottom. Then 10 kg each of SSP and gypsum and 5 kg urea is dissolved in water and the solution is sprinkled over trash followed by dung slurry (25 kg dung in 100 liters of water). After this a layer of half a ton of pressmud is spread uniformly. Whatever contents added so far is to be repeated couple of times till the pit completely fills. The top of the pit is then covered by plastering either with pressmud or with wet soil. If this is left for 3-4 months, one can get enriched compost to apply to cane field.

Pit method of enriched composting with microbial culture

In this method, a pit of any convenient size (as in above method) is dug and the bottom 15-20 cm is filled with dried trash. Over this, dung slurry and microbial inoculums / culture @ one kg per ton of dry trash are uniformly sprinkled all over the trash. Then to enrich the compost addition of 10 kg urea and SSP is necessary. This must be covered with soil to complete one layer. Likewise several layers are filled till the pit completely fills. The top of the pit is again covered with soil and watered to maintain optimum moisture in the pit for fast microbial decomposition and to get a completely decomposed compost in a period of 3-3 ½ months.

Heap method of enriched composting

Enriched composting can also be prepared above ground surface by placing the chopped dry trash on hard surface in any convenient size (5x4m length and width). The trash has to be pressed compactly to a thickness of 15 - 20 cm which is then sprinkled with dung slurry and microbial culture mixture. After this add 10 kg of N, P and K supplying inorganic to complete one layer. Such layers are to be repeated 3-4 times one over the other and top is completely plastered with mud. After 45 days the whole heap need to be mixed thoroughly turned upside down and compressed as before. This gives the compost in a period of 3-4 months.

In-situ method of composting

After the harvest of cane, the available trash may be composted in the field itself if the crop is ratooned. In the ratoon crop, the trash can be spread in the furrows and stubble - shaving and trimming the ridges can be done. Had the crop been harvested very close to the ground, stubble shaving can be skipped. 250 kg Mussoorie rock phosphate or SSP, 125 kg urea and 60 kg MOP/ha must be applied on the trash in the furrows and irrigation can be given as usual for the crop. The trash would be decomposed in six months. At this stage earthling up can be done for the crop.

When the trash is com posted in the field, weeds would be controlled. Soil moisture would be conserved and so irrigation frequency can be widened. This will be boon to the farmers particularly in drought prone areas. Because of soil moisture conservation, the plants can absorb the nutrients more efficiently and so the utilization of nutrients applied through fertilizers and the performance of the crop in terms of yield would be enhanced.

Experimental results have shown that the cane yield from the field in which the trash was composted *in - situ* was higher by 12-15 tonnes / acre than in the field in which the trash was burnt. We can take as many ratoons as we want as long as we manage to maintain the crop population at the desired level without the incidence of serious pest like scale insect or diseases like red rot. The overall productivity of the field can be enhanced through improvements in the fertility status, properties and health of the soil by composting the cane trash in the field itself.

Enriched FYM (E-FYM)

FYM is bulky and low in major plant nutrients such as N, P, and K. Hence there is a need to improve its quality. E-FYM is recommended to rainfed crops, which require available P for their root proliferation to withstand the initial growth stages under dry land conditions. First, 750 kg of well decomposed FYM is taken. After sieving, the recommended dose of P and K_2O for the crops to be grown is mixed with the sieved FYM. The mixture is spread in the form of heap and plastering is done with red-earth paste. This anaerobic process is maintained for 30 days. Then, the nitrogenous fertilizer recommended for basal dose of the crop to be grown is mixed and it should be applied immediately before sowing.

Weed Composting

Though weeds contain large quantity of nutrients, they affect the farm operations considerable; compete with crop plants for nutrients, space and water; poses health problem to human beings.

This problem can be overcome by bioconverting the weeds using lignolytic and celluloytic organisms like *Trichoderma viridie, Fusarium* sp., and *Pleurotus sajor caju* through composting which gives a "lean environment in addition to improved moisture conservation, water use efficiency, nutrient use efficiency and soil health. The weeds can be biodegraded individually or by pooling all the weeds together depending upon the quantity available and necessity. The arable and waste land weeds like Mexican poppy, thorn apple, nut grass, water grass, barnyard grass, crowfoot grass.- Wild grass, carpet weeds, Parthenium, Cassia etc and the aquatic weeds like Ipomea, water hyacinth and seaweed can be used for composting. A common bioconversion technology is suggested for all the weed materials. Composting of Parthenium weeds by this bioconversion technology is given below as an example to other weeds.

Composting of Parthenium

Parthenium, a problematic introduced weed prevailing in cultivable and fallow areas, affects crop yield and poses health hazards to human beings by causing skin and eye irritation, asthma, fever, etc. Parthenium seeds have long storage life and higher germination percentage and spread quickly by wind. For composting purpose, the Parthenium plants are to be cut into small pieces of 10cm size by using chaff cutter. These chopped materials are spread to a thickness of 10 cm. Over this inoculums viz., compost cultures *(Trichoderma viride, Fusarium* sp.) are spread uniformly. Then, weed material is spread over this to thickness of 10 cm. Over this layer, 0.5 per cent urea is spread (5 kg urea/t of chopped materials). This has to be repeated till about I m height is obtained. Then plastering with mud is done. Periodical sprinkling of water is done to maintain 50-60 per cent moisture. After 2 weeks, a through mixing is given. In 40-45 days, the parthenium compost is ready for field application. This compost contains 2.49 per cent N, 0.73 per cent P and 1.37 per cent K with a C: N ratio of 21: 1. The same method is followed for other weeds also. The Ipomea weed compost contains 2.49 per cent N, 0.32 per cent P and 0.50 per cent K with a C : N ratio of 20: 1 whereas the water hyacinth compost will have 2.11 per cent N, 0.3 per cent P and 0.73 per cent K with a C:N ratio of 22: 1.

Hints for Composting Agricultural Wastes Carbon : Nitrogen ratio decomposition process

Carbon nitrogen ratio is the relative proportion by weight of organic carbon nitrogen, in the soil or any organic matter. The number obtained by dividing the percentage of organic carbon by the percentage of nitrogen is usually referred to as carbon-nitrogen (C:N) ratio.

Carbon nitrogen ratio is of fundamental and practical importance in understanding the mineralization of the organic matter. It is a well established fact that the C:N ratio exerts a marked influence upon the mineralization of carbon or nitrogen of the green matter, both under aerobic conditions. Only green matter with C:N ratio of 30 : 1 or lower will decompose in the normal manner. Materials with very wide ratio do not decompose rapidly as the nitrogen contained in the green matter is not sufficient for the microbial activity and materials with very narrow ratio decompose more rapidly as excess nitrogen is available for the microbial activity. This is due to the fact that the various microorganisms taking part in the decomposition process require carbon for their energy and nitrogen for the synthesis of their protoplasm. Hence in the presence of more carbon (wide C:N ratio) more energy giving material will be available for the microorganisms and thus the increased activity of the organisms will utilize all the nitrogen. Once the available nitrogen is exhausted, the microorganisms will become inactive and decomposition rate will be retarded. In contrast, when proportionate amongst of both carbon and nitrogen are available, mineralization will proceed in a normal way.

Mature plants have a very wide C:N ratio of 50 : 1. In such cases, the carbon, when mineralized, gives out energy, which is utilized by the various microorganisms. With the availability of more energy, the organisms utilize all the available nitrogen present in the soil and plant material. Thus the entire nitrogen will be leach down and when this type of green matter is added to the soil, it will be acted only by the carbon mineralizing organisms. When matter with very narrow ratio, below 20 : 1 is applied to the soils, the availability of nitrogen will be more due to greater mineralization. However, in all cases of plant materials, finally the C:N ratio will be brought to about 10: 1 , which is said to be an equilibrium stage. Succulent and leafy portions of green manure, when applied, decompose very quickly (in about a week time) and behave almost like inorganic fertilizer. Thus, the C:N ration is an useful indicator by which the decomposition process, the release of nutrients and other biochemical chemical reactions connected with mineralization can be well understood.

From the several composting methods, the following few practical hints are offered for proper composting of agricultural wastes. Collect plants waste, and make the composting mass heterogeneous by mixing tow or their types of materials. Shred the material to 5-7 cm size. Mix the material with cow dung slurry in the ratio of 70:20: 10 (agricultural wastes: animal dung: soil). Add super phosphate or rock phosphate at 5 per cent (w/w). Maintain the medium moisture content at 50-60 per cent by sprinkling water. Do plastering with red earth or mud. Give periodical turning depending on the method employed. Within six months, the compost is ready for field application. If needed, retain a part to serve as mother and seed compost for subsequent cycle of composting.

How to produce fruit crops under organic farming system?

The fruit crops are becoming very important component in Indian agriculture. The different fruits form the nutritive part of human diet, being rich in vitamins and minerals. Besides carbohydrates and proteins, they supply good amount of different vitamins and minerals and minerals which are required for good human health. Therefore, they are called as protective foods. The Indian Council of Medical Research has recommended a balanced diet to content 90 g of fruits per day per capita. Whereas the present availability works out to only 55g of fruits per day per capita. It is therefore essential to increase the production of different fruit crops by adopting scientific management practices. The fruit crops are important for cultivation from the point of yield per unit area as they produce higher yield than field crops. Therefore, they are more remunerative to the farmers. Besides the aesthetic value of cultivation of fruit crops, the employment opportunities are more in the cultivation of fruit crops. The different fruits also provide, raw material for processing industries. During last few decades the export of different fruits and their products has also increased thereby providing more financial support to the farmers.

For increasing the production of fruit crops the plant nutrition play important role. The productivity of the crops depends upon the supply of nutrients in balanced manner. All the nutrients that are present in the soil may not he readily available for absorption by the plants. The availability and uptake of nutrients by the plant is influenced by several external, internal and plant factors such as climate, soil environment, root stocks, genotype etc.

The different plant nutrients are supplied through organic as well as inorganic sources. However, due to intensive cultivation and use of inorganic fertilizers more in proportion and organic sources less in proportion, the different problems are being faced by the farmers, like accelerated soil erosion. The microbial activities in the soil are also adversely affected due to more use of chemical fertilizers and pesticides. Heavy metals in soil and water resources are the principal causes of environmental concerns due to fertilizer use in agriculture. In order to improve the soil properties the use of organic manures is essential.

The concept of organic farming is to demonstrate the effectiveness of low cost agriculture thereby increasing the net income from successive crops and not only the structure and texture of soil but the quality of agricultural output. Organic farming has number of benefits like reduction in environmental pollution, control of soil erosion, improvement in structure and texture of soil, food safety, maintenance of biodiversity, animal welfare and most important is human health welfare. The major principles of organic farming are to use of organic manures to maintain the fertility of soil, use of bio-fertilizers, use of resistant varieties, better soil management practices like solarization, mulching, cover cropping, intercropping, crop rotations and use of pheromones.

A wide variety of organic products are sold worldwide and the demand for them is growing faster in Europe, USA. Japan and Australia. India is already exporting a wide range of organic products like tea, spices, cotton, rice, etc. Domestic market for them is status increasing. In recent years, as several thousand farmers switched over to organic farming, the organic production in India has considerably increased. Although no miracles can he expected from organic agriculture, yet it can be realistic and promising option, especially marginal and resource poor farmers.

Organic manures not only increase the yield but also improve physical, chemical and biological properties of soils which in turn improve fertility, productivity, water holding capacity of soil. Organic manures play an important role in quality production of crops. Now days, fruits produced with minimum or no use of inorganic fertilizers are preferred in export market. Also organic food fetches an impressive premium after 20 per cent higher than identical products produced on conventional or non-organic farm.

Cropping pattern

The crop rotation and intercropping is the key to success in organic farming. Cultivation of host plant with leguminous intercrop may increase the availability of nitrogen and ultimately increase the yield and improve the quality of fruits. Baghel et al. (1986) found that intercropping of bananas with mung, udid, soybean, groundnut or cowpea had significant adverse effect on banana yield and increased the net returns.

Vioayakumar et al. (1986) reported that planting of single or double row of *Leucaena* (subabul) hedges in coconut gardens meets the requirement of green maturing and mulching material in a alley cropping system.

Nawale (1998) and Ziauddin (2000) found the beneficial effects of organic farming in banana and sapota respectively for increasing the yield and quality of fruits.

Zhao et al. (2002) reported the technique of establishing the pollution free citrus orchard in china. In this technique the multi tier cropping system was followed from ecological benefit point of view (which helped to minimize insect pest and disease problems), 10 maintain soil fertility and improve the fruit quality. La Mondia (2002) studied the crop rotation pattern in strawberry by using Saia oat and Sudan grass. He found that the population of Leasion nematode was reduced and when the inter planting was followed the weed population was also reduced. The production of rotation crops suppressed the densities of pathogen, weeds and white grub.

Management of diseases and pest through organic farming

Pest, diseases and weeds cause an annual loss of about 30-40% potential food production in India. According to very conservative estimate the annual crop losses due to these pests amount to a staggering Rs. 25,000 cores. The current pest control technology heavily on pesticides. Due to the dangerous use of chemical pesticides, there scientist to look for alternative methods of pest control. The non-chemical methods of pest management certainly form the major components of organic farming. In organic farming systems pest control strategies are largely preventive, rather than reactive. Organic farming is not a single method but rather a variety of techniques which are aimed at reducing costs, preserving the environment and protecting human health by eliminating used of toxic form chemicals. Instead of using a chemical pesticide the bio-agents are use for control of pests. These are the plant extracts, which for the most time create unfavorable environment for pest by acting either as antifeedant or by disturbing life cycle. Thus the bio-agents rather support nonviolent farming. They can be easily grown and made at farm level. Some useful weeds first which is known one of the best bio-agents rather support nonviolent farming. They can be easily grown and made at farm level. Neem comes first which is known one of the best bio-agents. Some useful weeds like lantena; datura, tulsi, adathoda, etc. act as natural repellent to many pests. The indigenous trees like Daris, wood apple, anona and their bioproducts have excellent insecticide value in controlling diamond black moth, heliothis, whiteflies, leaf hopper and aphid infestation. The fish oil resin soap (FOS) is salf, non-poisonous, natural product used for the control of developed whiteflies, mealy bugs, wooly aphid, etc. These organic sources are more important in controlling the pests of fruit crops where the problem of pesticide residue is more serious, because the fruits (Vegetables also) are consumed in the fresh form immediately after their harvest.

How we can use organic farming for crop production?

Importance of low external investment sustainable agriculture (LEISA) technology does not emphasis Central Govt., State Govt., Agricultural

Universities, scientists, and eminent environmentalists are fully aware of multifacel hazards on animals, human beings, soil health there by posing problems. Policy makers are now busy in planning, taking into consideration the global challenges and international competition in respect of quality production of food, fruit, vegetable, spices, medicinal and other crops.

The wave has just now begun but unfortunately there is no precise package of practices for growing the crops with organic method. Krishi Vigyan Mandai, Nanded is NGO, active in this field under the guidance of honorable Advocate Manohar Parchure, Nagpur. This dedicated personality took us to the fields in Vidarbha, convinced and made us think and study deeply all aspects of organic farming and practice them on our respective fields. This made us study to evolve a common package of practices which a common cultivator can follow and start organic farming from coming season. We are confident that if our 10 points action plan is followed there is no fear of reduction in crop production, even in first year of switching over. We appeal to the scientists to take up extensive research work in every point of action plan.

1. Selection of proper crop/field

To begin with, one has to be careful in selecting proper crop suitable for particular field. Cultivator knows it very well that on which patch which crop grows well. Factors to be taken into consideration in selecting the crop and field are crop rotation (monocot/dicot crops, shallow rooted/deep rooted crops) previous crop, green manured crop etc. Single crop every year in particular plot must be avoided. Ideal rotations are for example: mung/udid *rabi* jowari/bajra/sunflower, gram, groundnut (JL 24) rabi jowar, cabbage, tur/soybean, wheat, groundnut sunflower etc. Mind that soil needs rest and green manuring off and on.

2. Use of straight / improved varieties

Green revolution phase introduced hybrid in every crop on a large scale. There is wrong impression in the minds of all farmers that hybrids are must for high production. But it is evident that our scientists from Agricultural Universities especially MAU, Parbhani have proved by figures that straight / improved verities are equivalent in crop production with hybrids. In case of sorghum PKV 801 (Sweta), PKV 809, swati, phule yashoda are giving satisfactory quantitative and better qualitative yields. In cotton crop NH 545. Turab, Anajali are also giving good yield with less plant protection 'costs. Vidnavardhini (MACS) Pune have released best varieties in soybean and wheat. Dr. Punjabrao Deshmukh Krishi Vidyapeeth, Akola has released many promising varieties in gram sunflower etc. Mahatma Phule Krishi Vidyapeeth, Rahuri has given wide range varieties in vegetable crops. Therefore, the farmers have many choices against hybrids. in organic farming hybrids must be avoided to save plant protection expenditure and to obtain quality production.

3. Seed / Set Treatment

Seed of cereals, pulses, oil seeds, vegetable or sets / planting materials / tubers Turmeric, potato, banana, sugarcane, ginger must be treated with *Azotobacter* + PSB + *Trichoderma viridae*. Whereas for dicot crops treatment with Rhizobium + PSB *Trichoderma viridae* is very useful for micropbial multiplication and protection agains fungal infection. In organic farming the use of *Triochoderma viridae* for every crop is invetible as it has proved very effective against serious fungal.

Farmers from Lingi, Barad, Jamner, Ajra (Dist. Kolhapur) are also impressed with the treatment of Rishi Krishi's material to seeds. *Amritpani* is the mixture of 1 kg cow dung, 25 g cows' ghee and 50 grams of honey with adequate water. It is supposed to help growth of micro organisms and act as attractant to the earthworms in the soil. This can be interesting aspect for scientists to study.

4. Sowing / Planting Method

One must be very careful in sowing / planting the crops. Because your method of sowing govern the yield, pest incidence and quality of the soil. Sowing of single crop is not advised in organic farming. In case of sowing of monocot crop it should be incorporated with the lines of dicot crop and vice versa. There should also be a mixture of few seeds having more height than main crop which would serve as birds peaches e.g. Maize in gram, jowar in wheat (nightmares to eat rats) etc, mixture of seeds should be considered with the object to supplement nutrients to take care of pest by biological control, to have symbiotic effect on the growth of main crop etc. Some of the examples are sorghum & Tur 3:3 or 4:2, Tur & Oilseeds 1 :3, Cotton & Soybean 1: 1, Sorghum/maize & soybean 1:2 or 4:2 maize & Tur 4: 1, soybean maize 3:2, cotton and Udid 4:2, etc.

Sowing should always be done North South to get magnetic effect of the poles and maximum sunlight on both sides of the plants. The practice of sowing dhaincha around or near sunflower reduces parrot attacks on sunflowers. Transplanting tomato, brinjal, okra on one side and coriander or radish on the other side will reduce pest infestations.

Latest integrated nutrient management of crops includes the sowing of cowpea (ten drilled variety) in rows of cotton, dibbling gram seeds in sugarcane crop. Mr. Suresh Desai R/ Bedkinal Dist. Belgaon (Karnataka) has evolved new method of sugarcane planting with the help of sugarcane bud scooper machine and plant them in furrows at 60 cm distance. The distance between two furrows is 360 cm (12 feet) Arogreen mixture of til, gram, methi and coriander is sown in between two rows of sugarcane. It is harvested 45 days after sowing kept in the rows as mulch. As per the opinion of Mr.Suresh Desai this biomass is capable of providing all required nutrients to the sugarcane crop. This fact has successfully been demonstrated in Belgaon, Jamner, Hoshangabad, Bedkinal etc. Vasantdada Sugarcane Research Station, Manjri, Pune is also undertaking experiments on planting of sugarcane by this method.

5. Use of Vermicompost

Every farmer has cattle on the farm. If he prepares vermicompost with the help of available dung and agriculture wastes, he need not purchase chemical fertilizers from the market. Crops grow with the help of microorganisms and not fertilizers. Earthworms are the managing director (MD) of all the micro organisms. They are capable of providing all essential macro and micro elements to the plants required at different stages of growth. Farmers can mix bio-fertilizers in vermicompost and drill the mixture at the time of sowing or hoeing the crops. It helps in improving yield in quantity & quality too. In general, paddy, wheat, sorghum, bajra maize, groundnut, Tur, vegetable crops need at least 500 kg vermicompost ha^{-1} cotton, banana, and sugarcane need 1000 kg of vermicompost ha^{-1} . There is no maximum limit of application. More you give, maximum you help to improve soil quality and obtain crop yield.

6. Use of *Amritpani*

There are controversies amongst scientists about the effect of Rishi Krishi's Amritpani which is the mixture of 10 kg of cow's dung, 250 gm of cow's ghee and 500 gm honey diluted in 200 liters of water and applied locally or through irrigation water to crops in 0.40 ha area. The crops of short duration need 3 application of Amritpani, whereas crops of long duration like sugarcane, banana 6 to 7 applications of Amritpani are recommended.

Advocate Manohar Parchure advises application of fermented mixture of 10 liter cat lie dung 10 its urine 100 g of jaggery with 200 liter of water to 0.40 hectares. It also helps to fulfill the nutrient requirements of the crops. We are also getting good results from the application of drive mixture to the crops.

7. Use of Dung water

Population of useful micro organism In the soil governs major role in crop production. It is revealed that there are maximum numbers of microbes in fresh cow dung as compared with FYM, as plant slurry. I appeal the microbiologists of MAU to examine and approve this fact. Our opinions are based on the results we are getting which may kindly be tested on scientific base. We have observed differences in treated and untreated plots. We therefore advise farmers to use their available cattle dung, dilute it with water and directly sprinkle or apply locally to the standing crops.

8. Precise Pest Management

Seven points mentioned earlier help one or other way to reduce pest infestation or induce pest tolerance. One must study and observe the critical economic threshold limits of every pest; and then only take pp)per action.

Scientists have already fixed these limits. Efforts should be made by all of us to educate the farmers and make them aware of the facts. Use of straight / improved varieties, mixed cropping, sowing of trap crops with main crop, keeps pest population naturally below economic threshold limits. If at all limit exceeds one can use NSKE, 5% cows urine seed treatment / drenching / dipping / spraying *Trichoderma viridae,* HaNPV, Tricho-cards, yellow colour painted tins (whiteflies) light traps (most effective for lemon butterfly on citrus). Vermiwash etc. are quite effective measures to keep pests under check.

9. Weed management

Late Prof. Dabholkar, Pune in his book vipulach srisht wrote a chapter "Tan deyi dhan". It is interesting to read the facts he pointed out. Usually it is taught to us by all that weeds are harmful and they are blamed as "Tan Khai Dhan" But nowadays scientists also agree that if weeds are managed they help us to increase crop yield Scientists and farmers are requested to study and observe the importance of every weed. Weeds do harm but at what stage must be learnt. Every crop has critical growth period, after which any amount of weed do not affect crop yield. In general we can say that weeds should be never let grow more than main crops height secondly 15 to 30 cm around plants base should be kept weed free. Early 2 hoeing / harrowing, 15 and 25 days after sowing help to control maximum weed population. Thereafter it is recommended to cut the grown up weeds including *Parthenium hysterophorus* etc. can be effectively used as mulching material or vermicomposting *Parrhenium sp* plant bears 6000 seeds which further spread with wind & water profuse growth of partheninum resulted after absorbing reasonable amount of solar energy which can be effectively utilized by organic farmers either as mulching the crops or use it for vermicomposting. Seeds of parthenium have no mortality, if at all they are dumped in compost pits but if they are placed on vermibed, earthworms digest all of them and convert into best vermicompost.

Many weeds act as repellants to pests. They exude hormones and enzymes from their roots, leaves, flowers which help in growth of plants many weeds exude chemicals which repel the harmful microbes & protect crop. Weeds of biodiversity and mulching of biomass, weeds, plant wastes etc. provide essential macro and micro elements required by the crops. Scientists have so far studied harmful effect of weeds on crops but they have not considered plus points of weeds which may be effectively used in organic farming techniques.

10. Mulching

Mulching the soil with agricultural wastes, dried leaves, sugarcane trash, dried wastes of turmeric, post harvest waste; weed material etc biomass creates congenial atmosphere in the soil for rapid multiplication of microorganisms. It keeps soil moisture, humidity, temperature ideal to the plant growth. Apart from conserving moisture it prohibits weed growth. It increases the organic carbon content in the soil. Mulching material is capable of absorbing atmospheric moisture and provides it to growing crops. In organic farming motherland is not supposed to be left naked but it must be clothed with biomass cover.

In our opinion above mentioned 10 points action plan help greatly the beginners who seriously think to shift over from chemical farming to organic farming. We admit that our statements may not be based on scientific base but the nature is the best teacher. Many mysteries are still unanswered by science.

Scientists are rich in knowledge and the farmers are rich in practical experiences. If both collaborate, sit together and share the facts with each other, a definite and concrete action plan can be prepared and presented to common farmer. Writer welcomes the opinion of the reader of this article.

Experiences in Organic Farming

Krishi Vidyan Mandal, Nanded is NGO active in Nanded district since 1996. Members of KVM are educated enthusiastic and progressive farmers. Many have been honoured as Shetimitra, Shetinishta, Udyan Pandit, Sinchanmitra, record yield achievers etc. KVM has Agricultural Library, Agriculture Technology and Information Centre (Internet), Mosambi utpadak sangh, Mahabanana sangh, self help group and organic farming project.

Organization has taken interest in organic farming from year 1997 when honourable and most respected advocate Manohar Parchure convinced the importance of organic farming to us. He insisted us to visit personally the organically grown crops in vidarbha region and then adopted the technology. We did so, got convinced and started learning and adopting the organic farming methods. The main problem before us was the package of practices of organic farming. We often visited each others' plots, discussed and decided the action plan. We regularly conduct monthly meeting and share our experiences. Our main crops are sugarcane, cotton, wheat, banana, tur, mung udid, turmeric, ground-nut & citrus.

Organic farming Technology cannot be explained in a nutshell as neither Agricultural Universities nor Department of Agriculture has given any precise guideline or package of practice right from land preparation to post harvest technology including marketing of organically grown food crops. We as per our understanding and experiences followed different methods to grow these crops. Records were kept to evaluate the practices. Experience was not at all encouraging in the beginning but the concept and its importance was very clear in the mind so we continued making slow progress.

Quality of organically produced food products

The modern organic farming concept is not new to this country. In fact the idea of organic farming emanated from the Indian experience. The demand for organic food has grown fastest in countries which used the maximum of chemical like fertilizer, pesticide weedicides, growth regulators, antibiotics etc. as awareness grew of their ill effects on human health and soil. In India sixty per cent of our cereals, millets, pulses, oilseeds come from areas where little, if any fertilizer or pesticides are used. Before the advent of fertilizer and pesticide use in late 1940s and early 1950s in India all agricultural produce was organic. They were grown with the help of organic manure and plant protection as and when needed was done with indigenous methods using no synthetic chemicals. Biodiversity of soil was maintained with appropriate cropping pattern and mixed cropping. Agriculture as a way of life with animal husbandry was an integral part of the system. The health benefits of organic foods are: chromium is found in organic food and essential for normal utilization of glucose. Selenium an antioxidant nutrient that protects against cancer and heart disease is also higher in organic foods, calcium, boron, lithium, iron, vitamin C are all found more in organic foods compared to ordinary foods (Jha, 2003).

Organic farming has become a subject of debate nowadays and supporters of organic farming claim a lot about benefits of the system without adequate research support. Here, an attempt has been made to compile available information on impact of organic farming on quality of the product. It is universally agreed fact by the crop nutritionists that chemical fertilizer acts in exactly the same way as nutrient from organic sources in the soil, since they are chemically same. It is also true that the quality of the agricultural produce, particularly horticultural produce like flowers , vegetables and fruits improves when the nutrients are supplied through organic manures than in the form of fertilizers. This is because of the supply of all the growth principles like enzymes, hormones, growth regulators etc besides all the essential plant nutrients by the manures. As a result, the metabolic functions in the plant get regulated more effectively resulting in better synthesis of proximate constituents and consequent improvement in the quality of the produce. Fertilizers would supply one or a few nutrients only, and not the growth principles like enzymes and growth regulators (Kumarswamy, 2002). Further, organic fertilization practice may be more resource efficient especially in terms of reserves of (recoverable) nutrient elements. According to Sankaran (1996) organic manures in comparison to inorganics are superior and beneficial because of nutrient dynamics, the soil structure buildup and stability, optimum air water relationship, retention and release of nutrients and regulations are all favoured with organics, this creates balanced supply of nutrients to all metabolic functions in plants.

Quality of food, feed and fiber refers to the value which is attached to the product with respect to its properties, namely nutritional properties, hygienic properties, organoleptic properties, functional properties and environmental compatibility (Krauss, 2001).

Urbanization is one of the major driving forces behind production of quality crops. Urbanites, usually having a better income than their rural counterpart and demand more and diverse food such as fruits and vegetables. This benefits those producers who can supply quality food accordingly. Transferring food into urban centers means also transfer of nutrients from the field into the town. Replenishment of these nutrients in the form of sewage slurry is hardly done because of fear of possible contamination with heavy metals and undesirable or even toxic substances. The export of some 110 million tons of fruits and vegetables from South Asia represents around 650,000 t N, 160,000 t of P_2O_5 and 375,000 t of K_2O and to meet this requirement it is insufficient to rely on native soil fertility. Use of organic manure or recycling of plant residues are not in a position to compensate for nutrient losses due to transfer of food into town or export. Sustainable crop production with management of soil health and quality of the produce is prime importance for the agricultural scientists, planners and farmers in our country.

Food quality

According to Abalaka (1999) "an intrinsic property of food by which it meets pre defined standard requirements. Determinants of food quality can be grouped into several properties. Food quality therefore refers to the value, which is subjectively or objectively attached to food with respect to quality properties".

Hardter and Krauss (1999) also added the aspects of health, sensorial, suitability, sociophysiology, political-societal and ecological values.

Nutritional properties includes chemical composition of food commodity i.e. carbohydrates, proteins, oils, minerals, dietary fiber, vitamins etc. whereas, functional properties includes processing quality, storability etc and organoleptic attributes are in concern with taste, colour, flavour and consumers acceptability. However, hygienic properties include incidence of pest and diseases, residues of the pesticides, microbial and mold contamination etc. Moreover, in environmental compatibility the consumer, especially in developed countries become more suspicious on how the food had been produced. He not only looks for affordable food in good quality but also for 'safe' food. He wants to follow-up the production, whether environmental friendly (Krauss, 2001).

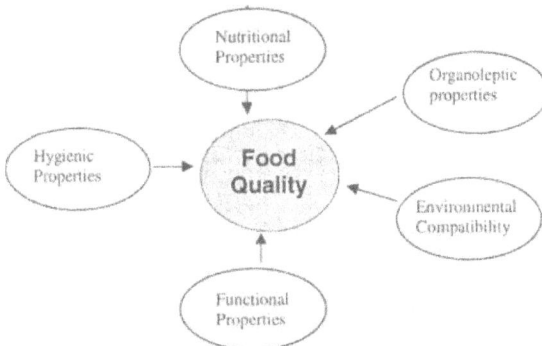

Most of the progressive organic farmers are of an opinion that the highest quality food is grown in the healthy land in a natural ecosystem. Nirmala (2003) in her review related to food quality of organic farming mentioned that the research in this regard is inconclusive and needs further detailed investigation. But, Sharma (2001) cited that food produced using organic methods tastes better; contain a balance of vitamins and minerals than conventionally grown food. However, the scientific evidences in this regard are meager with some studies showing increase in vitamin C , minerals in fruits and vegetables and protein in pulses . The vitamins and mineral content of food is controlled by a complex interaction of factors including soil type and ratios of minerals in soil, added compost, manures and fertilizers. It is therefore difficult to separate the influences of environment and farming system:

Holistic quality methods, such as picture developing can be used successfully to distinguish the crop produced from different farming systems. As per report organic wheat has lower protein level than conventionally grown crop. However, milling quality in organic wheat can be achieved through careful choice of variety and management.

Organic verses other farming systems in relation to production and quality of produce. Many claims related to improvement in shelf life and quality of the produce have not supported by scientific data and warranted for an in depth analysis to provide a sound basis of the same. Majority of the soil scientists and crop nutritionist are of the opinion that integrated nutrient management using manures, fertilizers and bio-fertilizers will facilitate restoration, improvement and maintenance of soil health, which will guarantee agricultural production at high levels with high quality produce as well. Agriculture at high level of productivity could be sustainable only through such integrated way. It will also safeguard the environment and natural resources. The philosophy of sustainable agriculture will become a bitter irony, if fertilizer use is reduced or excluded in the name of quality improvement of produce or environment protection. Though indiscriminate use of fertilizers and chemicals has a potential for polluting the environment, it does not mean exclusion of fertilizer altogether. Intensive agriculture on commercial scale cannot be sustained for long through total organic farming (Kumarswamy, 2002). As per Chhonkar (2002) organic farming will not be able to sustain food production at levels to meet the needs of food and fiber for our burgeoning population. What is the productivity of the system and what level of yield is acceptable to us? Nobel Laureate Dr. Norman E. Borlaug has to say *"We can use all the organic that is available but we are not going to feed six billion people with organic fertilizer"* (Borlaug, 2002). Very recently, Chhonkar (2003) cautioned that there are several myths propagated by a dedicated band of chaemophobes about organic farming and organically produced food. Few of them are given here.

Organic food tastes better and is of superior quality

The traditional belief that organic manure promotes quality while mineral fertilizers promote quantity was shown to be over simplistic by Schuphan (1974) on the basis of trials conducted for over a decade. Regardless of whether the nutrients are from organic or inorganic source, plants absorb the same in the form of inorganic ions: ammonium, nitrate, phosphate, potassium etc. Sensors in plant roots if any, to distinguish between nutrient ions coming from organic or inorganic source have to be still discovered. Once absorbed the nutrients are resynthesized into compounds which determine the quality of produce e.g. flavour, shelf-life etc., which is the function of genetic makeup of the plants (variety). There is no scientific evidence presented as yet to show that organically produced food is of better quality and taste, and use of chemical fertilizers deteriorates it. The better taste of the organically grown food is of psychological in nature and could be attributed to 'Placebo effect' widely used in drug testing, where harmless sugar pills administered to control groups are known to cure patients of their imaginary ailments, when told of its novelty and wonderful therapeutic properties. More someone pays for it faster is the cure: a clear case of 'mind over matter'.

Organic food is more nutritious and safer

There is a general perception in public minds that organically grown food is more nutritious, healthy and safe. There are no consistent and valid reports of differences in the mineral contents of organic and conventional food. However, N applications generally improve both the protein and bread making quality. There are many factors, environmental and cultural, that influence the nutritional composition of the produce. It is at best confusing to give credit for these changes to organic cultivation. There is no difference between the protein content and other quality parameters such as vitamins, nutraceuticals and trace minerals of conventionally and organically grown crops which at best could be linked to the varietal characteristics. The genetically modified 'yellow rice' which was in the news recently owing to its higher vitamin A content over the traditional varieties will continue to have its superior nutritive value irrespective of organic or inorganic fertilization. In the field of plant nutrition the cry of 'only natural' has no justification or scientific basis (Woese *et al.* 1997). The attitude that organic foods are safe and healthy is based on misconception that hazardous is foods are mainly derived from agro-chemicals are the major source of the food-borne diseases such as typhoid, gastroenteritis, dysentery, cystecurcosis, etc. Animal wastes contain intestinal bacteria, many of which may present substantial human health threats. Land application of manure is particularly associated with *Salmonella*, *Escherichia coli* and *Taenia soleum* which can contaminate the soil. These pathogens are known to survive in soil for a long period. They may be carried on edible plant parts coming in direct contact with soil and get into the food chain. They may also be introduced into shallow surface waters as well as ground water polluting potable water supply (Mikkelsen and Gilliam 1995).

Results emerged out from long, term fertilizer experiments (LTFE) started in 1970 clearly indicated that use of fertilizer along with organic manures is only sustainable for production and nutritional security (Goswami, 1998).The data pertaining to the effect of organic farming on quality of the produce in literature is insufficient to draw any conclusion. However, many of the research workers evaluated quality attributes of different agricultural commodities in other farming systems which are being cited here.

Chen *et al.* (1997) observed an increase is sugar and juice content in the fruits of citrus in response of chemical fertilizer plus dolomite lime plus organic fertilizer. Whereas, Huchche *et al.* (1998) observed increase in shelf life and quality of citrus with organics as compared to inorganic applications. Further, Malewar *et al.* (1998) also reported increase in the yield and vitamin C content of chilli with application of N through urea and FYM (50:50). Further, Hangarge *et al. (2002)* found significant increase in length and diameter of chilli with organics in the form of vermicompost @ 5 t/ ha + organic booster 1 lit/m^2 and soil conditioner @ 2.5 t/ha + organic booster treatment over recommended dose of fertilizers (RDF) alone. Fruit quality parameters viz., total soluble solids: titrable acidity (TSS : TA), ratio of citrus juice, index of peel colour, the percentage of mesocarp and specific gravity of fruits were higher in the -OM treatment than in the + OM treatment (Tachibana and Yahata, 1998). Ingle *et al.* (2001) reported maximum yield with appreciable fruit quality of *khasi mandrill* from balance nutrition through combination of organic (neem cake) and inorganic fertilizers. In contrast, Sharma (1995) found significantly higher pod yields of Okra and its protein content with 10 tonnones compost + fermented cow dung as compared to recommended NPK in Vertisols.

Amar Chandra (1998) reviewed the work carried out with reference to yield and quality of vegetables in Madhya Pradesh and noted that biofertilizers along with inorganic N application was superior over inorganic Nor biofertilizer alone in increasing yield and quality of okra, brinjal, tomato, French bean and cabbage. Shreshtha *et al.* (1996) reported that inoculation with VAM fungi improved the fruit quality of Satsuma Mandarin with reference to Hunter's a/ b value of peel colour and juice sugar content.

In cotton the increase in seed cotton yield with the application of green manure to supply the 40 kg N/ ha was recorded. Similarly. quality parameters i.e. mean halo length, ginning outturn, seed and lint index was also found significantly superior in same treatment as compared to inorganic N (40 kg/ha) through urea (Chitdeshwari and Pravikesavan, 1998). Malewar *et al.* (2001) also reported increase in the sugar content of banana *(Ardhapllri)* with recommended NPK but it was further improved signil"icantly when organic booster 6 lit/plant or vermicompost (5 t/ha) was added.

International Federation of Organic Agriculture Movement (IFOAM) have defined some rules for organic farming and on that basis the importing country is verifying the certificate of organically grown commodity obtained by licensed exporter. Before import on the name of quality only pesticide residues are checked, if these are as per norms green signal is given for import of that food commodity. Here the particular quality criteria are only hygienic property of the food: nutritional, organoleptic and functional properties are not being evaluated for organic foods by the developed countries also. This means peoples are more conscious about toxic ingredients rather than nutritional, functional or organoleptic attributes.

Conclusion

Quality of the agricultural produce improved with organics because of the supply of all the growth principles like enzymes, hormones, growth regulators, etc besides essential plant nutrients but it does not mean that they can substitute the inorganic fertilizers because lot of data is available to prove quality enhancement in integrated nutrient management (INM) rather than organic farming. The data pertaining to quality of organically grown food in literature is very meager except pesticide residues and vitamin C content. Hence, needs in-depth studies by other human nutritionists.

Different problems and limitations in organic farming

Organic farming has been sustainable over centuries in India and other countries however; interest expertise and modern research backup in it are neglected totally in recent times because of success of modern farming, technologies, which are highly input intensive, resource deteriorating and environment polluting. With the increase in fertilizer prices due to decontrol, indiscriminate use of pesticides, increasing resistance in pests against pesticides and unplanned use of irrigation water have threatened to the sustainability of agriculture production, human and animal health hazards and polluting soil, water and environment. These likely problems have increased the relevance of developing alternative farming system in harmony with nature. The present input farming system arc considerably Jittered in concept, objective and components with organic farming / agriculture. The main feature of organic farming are (1) minimum tillage, (2) use of well decomposed organic manure, (3) recycling of organic residues, (4) proper crop rotations, intercropping, mixed cropping and polycropping, (5) use of biofertilizer, (6) mulching of weeds, (7) integrated pest management or no use of chemical insecticides (Bhole, 1992; Baphna, 1992; Dixit, 1993.)

Controversial Views

There are no two thoughts about the beneficial effect of organic manure's on soil biology, soil fertility, release and availability of nutrient 10 crops (Reddy *et al,* 2001). Shinde & Gawade 1992, Malewar et al; 1999; Shinde, 1992; Mishra *et al.* 1992 & Malewar, 2004). However, it seems that there is lack of correct understanding of the organic farming concept probably because of inadequate research database. Many farmers believe that organic farming means only replacement of chemical fertilizers, pesticides, weedicides and insecticides by organic sources of nutrients and plant derived indigenous pesticides. According to Zende (1998), organic farming in reality aims at strengthening the ecological base of farming comprising soil and water resources. Dixit (1993) further supported that it is the method of farming system which primarily aimed at cultivating land and raising crop in such a way as to keep the soil alive and in good health. In rainfed farms, farmers are invariably using organic manure as plant nutrient sources. They are of the view that chemical fertilizers deteriorate soil health and resulting poor grain quality. The prohibiting cost of inorganics, risk of crop, failure, low purchasing power and ignorance of rained farmers are largely responsible for such a myth. (Sarkar, 2000), Dr. R. S. Paroda has emphasized this issue in his inaugural address at the 64th Annual convention of ISSS at Coimbatore. He has cautioned about increased risk of leaching losses and slow release of nitrogen during crop growth period with organic manures. Though ,many agencies are advocating ecological farming, many of such contentions do not have a scientific rationale, /research on crop yield, profitability, sustainability with organic manure or balanced fertilizer use over the four decades in permanent manurial trials of Ranchi has shown that: (I) in acid soils of eastern India, the best management practice (BMP)is lime + NPK application in crops (ii) Organic manuring produces crop yields that are about one half to two third of those obtained with lime + NPK (iii) Soil analysis reveals that organic manured plots, though improved soil properties considerably over the years, could not sustain high crop yields due probable) to its slow release of plant nutrients and comparatively lower impacts on exchange acidity, soil pH, exchangeable C and P status of acid soil .(iv) In HYV'S of crops, lime + NPK was always more responsive compared to organic manured Plots and(v) the sustainable yield index of organically manured plots was 0.30 for maize and 0.23 for wheat compared to 0.58 for maize and 0.47 for wheat in NPK + lime plots.

In presidential address of ISSS, Chhonkar (2001) cautioned about organic and biodynamic farming, a climate is being slowly and steadily built up in the country against the use of pesticides and fertilizer, favouring production of the so called pure foods, The protagonists of organic farming are spearheading this thinking much of which is without any scientific basis. He said we cannot ignore the presence of a very active and vociferous lobby which rightly or wrongly believes that some of the publicly funded research has dangerous implications to the country. In this soil scientists at present are not in frontline facing the flask as harshly as the biotechnologists are but we cannot be far behind because of our involvement in fertilizer and pesticide use research. Many people complain of loss of quality of produce due to use of chemical fertilizers. This loss of quality is rather hard to determine and is more of psychological in nature. There could be soil factors which have implication in the quality of produce such as flavor, keeping quality etc. There is no place for superstitions in science. Biodynamic farming is being propagated in the name of environmental protection and exploiting religious sentiments.

In nineties, a committee of central agriculture and cooperation, department, ministry of agriculture was constituted to investigate the impact of organic farming and its usefulness in India. The observations and comments of this committee are: (i) consider the food salivation of present and future, complete stop in use of chemical fertilizers in Indian agriculture would adversely affect the agricultural production and it will be difficult to achieve the food production target. Thus , use of chemical fertilizers may be reduced slowly and increase the use of organizes and bio-fertilizers, (ii) There are no two thoughts in advocating integrated pest management (iii) No consistency in implementing organic farming was noticed except at 1-2 center (iv) There is need of scientific footing for organic farming which is neglected so far.(v) No adequate infrastructure and facilities are developed in state agricultural universities for research in minimum use of inorganic and maximum use of organic,(vi) Quality of organic resource is many times remain doubtful further no control on marketing of industrial organic manure. (vii) No guidelines are available for organic farming policy, organic farming act, establishment of organic farms, certification, marketing of organic produce and standards etc. These observation, thus, suggest the need to critically analyze the limitations and constraints in organic farming under Indian conditions

India is divided in to 20 agro-ecological zones, varied in soils, climate complex, cropping patterns and total growing period of crops, Major part of the country comes under dry and hot belt with monsoonic type of rainfall, The rainfall intensity, total rainfall, distribution pattern varied greatly with dominance received during June to September with one or two dry spell of 15 to 20 days. All states of eastern part of the country, Uttarpradesh, Uttaranchal, Himachal, Jammu-Kashmir, coastal belt of Kerala, Karnataka, Goa , Maharashtra and the Andaman & Nicobar island receive rainfall more than 2500mm, Plains get rainfall in between 750 to 1250mm, However, North-west part of country and arid Rajasthan receive only 125 to 250 mm rainfall in a year. Temperature differences are also considerable within the various states, In general, overall climate is dry and hot (arid to semiarid) and hot semi-humid and humid except during rainy months and mountainous regions, nearly 30% area is under irrigation, the climate showed change wherever perennial irrigation is available, remaining 70% area is either rainfed or dryland. Although, in extent carbon loss is widespread from vegetative denuded soils across all the environments it is far more virulent in dry land under the influence of arid, semiarid and sub-humid tracts, majority of tropical Indian soils which belongs to arid and semi-arid climate, rarely, exhibit organic carbon levels exceeding 0.6% (Virmani *et al.*, 1982) and more commonly 0.5 % organic carbon. Jenny and Raychaudhary (1960) Outlined an inverse relationship between soil organic matter content and mean annual temperature, the rate of decomposition is expected to increase at least by a factor of two with every loose rise in mean annual temperature (Van't Hoff's temperature rule) Kayla and Sharma (1991) and Waikar *et al* (2003) described variations in soil organic matter with reference to climate. Organic matter distribution across soil was influenced strongly by mean annual rainfall, representative for a region, (Katyal, *et al*, 2000 and Waikar *et al*, 2003) it's content fell with decreasing rainfall and majority of soils showed organic carbon content less than 0.5%. Further, Jones (1973) reported a decrease of 0.17% in organic matter per 100 mm decrease in main annual rainfall in the Savannah region of northern Nigeria. Therefore, maintenance of organic matter particularly in tropical soil warrants for more concerted efforts than temperate soils. Lal and Kang (1982), while reviewing the organic matter management in tropical soils cited finding from previous studies that confirm necessity for at least 4 times higher organic matter input in tropical than in temperate environment to maintain the native .level of organic matter. Buyanovsky and Wagner (1998) described lower' equilibrium and upper equilibrium limit of organic matter. Tropical conditions favour organic matter decline because of low or no organic matter additions, accelerated degradation and loss due to year round prevalence (}f biological active temperature and moisture regimes as a result, rapid reduction in soil organic matter take place and the organic matter equilibrium tilts towards lower limit. Similarly over cultivation tends to attain the lower equilibrium; However, organic matter additions and fertilization tilt it in the direction of either new dynamic equilibrium or upper equilibrium ceiling. Thus, maintenance of desired (native) level of organic carbon in Indian tropical soil is the major constraint for organic farming. No standard of Indian organic agriculture are developed for soils, their organic carbon thresh hold values, soil fertility and soil management. Eswaran and Virmani (1990), while describing soil component in sustainable agriculture in semi arid tropics indicated 1.0% soil organic carbon as essential criteria for sustainable productivity. Similarly, other aspects related to carbon management in term of C:N, CP, C:S and C:N:P:S ratio and overall quality of soil organic matter are lacking in manurial policy (Malewar, 2004).

Though, appropriate soil management is fundamental to successful organic production, due allowance to this primary aspects is not of in organic farming in reality, the development and protection optimum soil structure and fertility, is the main goal of soil management.

AIFOF (1996) described broadly management which should ensure; regular input of organic residues, a level of microbial activity sufficient to initiate the decay of organic material, condition which ensure the continual activity of earthworms and other soil stabilizing agents, as far as possible, a protective covering of vegetation, deep loosening of the soil, minimal disruption of top soil and timeliness to ensure appropriate tilth and to avoid damage to existing structure.

Crop rotation should be as varied as possible and should aim at : maintaining soil fertility, reducing water logging and nitrate leaching, reduce weed, pest arid diseases (IFOAM, 1996). However, the use of engineered seeds, transgenic plant or plant material is restricted. Continues cropping of onions, mustards, cabbage, cauliflower and potatoes in the same plot in successive years is prohibited. Similarly continuous cereal rotation is also prohibited (AIFOF, 1996). Continuous potion of the same cropping system like rice-wheat in Indogenetic plains year after year poses constraint to the maintenance of soil organic matter to the expected level (Manna, 2002).

The availability and organic resources management is another constraint in organic farming. Tandon (1997) estimated nearly 700 million tonnes potential availability of organic materials. However, in reality, only a fraction of this is available for actual field application. Same author place the potential at around 30% of the total availability. Thus, nutrients available estimated are 5.65, 6.24 and 7.75 MT NPK during 2000, 2010 and 2025, respectively and probably could not meet the requirement of crops only by organic resources.

Growing industrialization and urbanization is generating solid waste, sewage-sludge effluents, agro--industrial waste and city garbage may be the alternative for organic agriculture. However, sewage and sewage sludge are prohibited / banned in organic farming (IIRD, 2001). Thus, city garbage and industrial waste are restricted / limited.

In addition to above, difficulties in collection and consolidation of organic resource, unavailability of research backup in identifying and using decomposing/ degrading micro biota, more times for FYM and composting, untrained human resource and uneducated farmers are the associated constraints in organic farming (Malewar, 2004).

Biofertilizers in Organic Farming

Biofertilizers have been recognized as vital component of organic farming. They are renewable eco--friendly source capable of rendering unavailable source of elemental N, bound phosphate, micronutrient and decomposed plant residence in to available forms in order to facilitate the plants to absorb the nutrient. Micro-organism which could be used as biofertilizers include bacteria, fungi and blue green algae. The inoculums carrying these organisms need to be applied to enhance their activity in the rhizosphere.

Rhizobium, Azotobacter, Azospirillum, blue-green algae. Azolla, phosphate solubilizer and mycorrhizae are the major biofertilizers available in Indian Agriculture and BNF process does not cause environmental pollution and renewable (Shankaran, 1993).

Among biofertilizers N fixing bacteria (Rhizobium, Azotobacter), Blue green algae, PSB and mycorrhizae play significant role in pulses, oilseed, cereals vegetables, sugarcane etc. The largest contribution of BNF to agriculture is derived from the symbiosis between legumes and species of Rhizobium. Inoculation of pulses is long established and successful practice to ensure adequate N nutrition in place of fertilizer N in most of the soil. In most of the studies on biofertilizer, there is saving of fertilizer N to the tune of 15-25 Kg N per hector (Dixit et al, 1992; Patil et al. 1992; Raut et al. 1995; Agarkar, 2002, Singh 1993). Many common heterotrophic soil bacteria and fungi are able to bring sparingly soluble / insoluble, inorganic / organic phosphate into soluble forms by secreting organic acids that lower soil pH and in turn bring about dissolution of immobile forms of soil phosphate. Some of the hydroxy acids may chelate with Ca, AI, Fe and Mg resulting in effective solubilization of soil P. Recently, several strength of P solublizing bacteria and fungi have been isolated. Field studies carried out in India have shown significant effect of seed / soil inoculation with PSB in chickpea, pigeon pea, soybean, and lentil. Increase in grain yield of chickpea, increase in P uptake and even protein content was observed due to inoculation of PSB (Ghonsikar and Shinde, 1997). In a compilation, Asalmol (2002) noticed that YAM, apart from N and P, also make available K, Mn and other micronutrient to the host crops. It may be possible to integrate the use of more than one biofertilizer and organic manures for increasing crop yield (Algawadi and Kulkarni, 1993). In India, where mix cropping and double cropping followed in many places, the combined application of more than one biofertilizers holds great promise.

In fact, biofertilizers and organic manure are complimentary to each other and co-existed. However, large -numbers of constraints were identified (Singh, 1993; Shankaran, 1993). Even though, field-tested, efficient inoculate strains are available, there is tendency of manufacturers to use strains of their own for mass multiplication and distribution to farmers. General experience shows that any singles suitable for whole country is rarely available at present. Singh (1993) identified several constraints, which are responsible for unsatisfactory results of biofertilizers. These are (a) biological constraints (i) presence of native ineffective strains.(ii) presence of antagonists which minimize the number of BNF organisms, (b) technical constraints: (i) mutation during fermentation, (ii) lack of soil specific strains, (iii) inadequate shelf life; (iv) competitive strains (c). Environmental constraints: (i) poor crop management (ii) poor water management (iii) soil conditions which limit N fixation, (iv) variation in climate and temperature. (d) Marketing constraints, (i) Farmers unawareness about biofertilizers (ii) Inadequate distribution system (iii) In adequate instructions (iv) Lack of technical support. (v) Wrong marketing practices (vi) Deficient quality of products (vii) Market segmentation, (viii) Seasonal demand.

Despite tremendous potential for use of various biofertilizers under different soil-climatic conditions and crops, the large scale use of biofertilizers has not yet become popular (Mostsara, 1993). He recognized the constraints and limitations as (i) Production and 'supply of often poor quality biofertilizer, (ii) Inadequate quality control facilities as production level and also at the distribution level, (iii) Short shelf life of biofcrtilizers necessitating their use within a limited period of 40 to 50 days time after production, (iv) Improper storage and transportation facility resulting in to deterioration of quality of product, (v) Lack of suitable marketing system and unawareness of dealers about the usefulness of biofertilizers and (vi) Unawareness among farmers and extension workers about potentials of biofertilizers and their use.

Green Manuring in organic farming

Green manuring and / or inclusion of short duration legume in crop rotation is a usual component of organic farming. Cultivation of legume in the field and their incorporation in to the soil can serve as alternative source of nitrogen and organic biomass C, Green Manuring with legume enriched soil N due to fixation of atmospheric N. The decomposing green manure has solubilizing effect on NPK and micronutrient in the soil. It also reduce leaching and gaseous loss of N. Depending up on a leguminous crop, it produces 10-25 tonnes of green matter per ha which will add about 40 to 90 kg N ha. Green leaf manure refers to turning under of green leaves and tender green twigs collected from shrubs and trees grown on the bunds, waste land and nearby forest areas. Common green manuring crops are Dhaincha, Sunhemp, Sesbania (*Sesbania* species, *S. rostrata*). However, common shrubs and trees useful for this purpose are karanj (*Pongamia glabra*), Glyricidia (*Glyricidia moculeata*), Neem (*Azadirachta indica*) and Subabul (*Leucaena lencephala*). Leucaena when planted as alleys with rabi sorghum and lopping used as manure added 37.6 kg N ha to the soil. The net effect of this input on sorghum yields was equivalent to that from the application of 25 kg N/ha through fertilizer (Patil and Kulkarni, 1988). Similarly, intercropping of pearl millet legume also found to help in reducing N requirement to cotton to the tune of 25% (Jadhav et al. 1992). In spite of many advantages of green manuring for supply of nutrients, organic biomass C and N and improving in soil physical properties, this practice could not popularize in farmers because of limitations. In many cases, there is loss of one seasonal crop due to adoption of green manuring practice; secondly, the problem with green manuring crop is thus they compete with cash crops for space, time, water and other inputs.

Recycling of crop residues

It is assumed that 50% of the waste material finds its way back to farm. The potential contribution of agricultural crops residues, rural and urban waste all together could he 4.4, 6.4 and 7.2 million tonnes of N, P, 0, and K,O in the year 2011.2031 and 2051 A.D, respectively (Sekhon, 1994). Thus, this precious little organic source of nutrients which would at the best constitute 15 to 20 of the total NPK requirement he used with advantage to protect the soils non-economic plant parts that are left in the field after harvest and remains that are generated from packing sheds or that are discarded during crop processing. The greatest potential as a biomass resources appears to be from the field residues of corn, wheat, soybean, and grain sorghum and the processing residues of sugarcane (bagasse) and cotton (gin waste and stubbles). In addition to above, process 'waste like groundnut shell, sunflower hull, rice husk, coir pith and cobs of maize, bajra, sunflower, and cotton gin waste. The residues can directly apply to the field and ploughed in. One way of profitability recycling the crop residue in their use as surface mulching materials. Mulches are thermo insulators and barrier to vapour transfer and can' also be used for moisture conservation. Incorporation of short duration *kharif* legumes green gram, black gram and soybean increased not only yield but also made available nutrients to succeeding crops (Shinde *et al.* 1966). These crop residues and on farm wastes are not properly utilized. According to the estimates, 50% of livestock dung is used for preparing dung cakes as fuel. Similarly, the crop residues/ processing residues are used as animal feed and other domestic purposes like fencing and roofing.

Issues and strategies for organic farming

1. Though all the agro-ecological regions are not suitable for organic farming in India, humid and cold belts and coastal strips can be very well utilized for organic farming. Thus, there is need to develop soil-climate - crop specific areas for organic farming.

2. National estimates of organic resource of nearly 700 million tones need to be consolidated biologically / manorial viable/ usable. Strategy to use cow dung / livestock dung in biogas technology for dual purpose energy fuel and biogas slurry manure may be strengthened by restructuring subsidy and training to farmers for developing agricultural skills for efficient working of biogas plants.

3. There is need to revert back to integrated farming, Crop- livestock, crop-livestock-forestry, crop- poultry-fishery, crop -livestock-horticulture and / or crop-livestock-forestry- poultry-fishery farming system may be developed. The strategies may be developed for integrated farming research in understanding linkages and complimentary of different farm enterprises and bioconversion process of farm waste.

4. Crop residue management and conservation tillage are on positive side of organic farming. Thus, strategies to recycle crop residues more efficiently are to be explored.

5. Macro and micro biota efficient strains locally available required being isolated for developing biofertilizers which is a component of organic farming. New frontiers in area of soil microbiology and biotechnology with reference to sulphur, zinc and iron bacteria need to be intensified.

6. No appropriate and optimum soil management practices are well defined including maintenance of optimum level of organic carbon, according to

varied agro ecological zones. Thus, strategic research on optimum soil management practices and maintenance of soil organic matter need to be undertaken. In addition, there is need to prescribe the nature and quality of organic residues/ organic matter/ humus for organic farming.

7. Various NGO's framed approximate standards for soil preparation to production of organic food/ fruits/ vegetables. However, there is need to have uniform standards for international market / Indian market separately.

7

FORMULAE AND EQUATIONS

1. Formula of Sokes's Law

$$V = \frac{2}{g}\frac{(dp-d)gr^2}{n}$$

Where,

V = velocity of fall (cm/sec^2)

G = acceleration due to gravity (cm/sec^2)

dp = density of the particle (g/cc)

d = density of the liquid (g/cc)

r = radius of the particle (cm)

n = absolute viscosity of the liquid (poise or m poise)

2. Density

$$\text{Density}(D) = \frac{\text{Mass}(M)}{\text{Volume}(V)}\text{gm/cc } or \ \lambda b/dft \ or \ \text{Mgm}^{-3}$$

3. Relationship between porosity and densities of soil

$$\% \text{ solid space} = \frac{\text{Bulk density}}{\text{particle density}} \times 100$$

since,   % pore space + % solid space = 100

or   % pore space = 100 – % solid space

or $$\% \text{ pore space} = \frac{100 - \text{bulk density}}{\text{Particle density}} \times 100$$

or $$\% \text{ pore space} = 100\left(1 - \frac{100 - \text{bulk density}}{\text{particle density}} \times 100\right)$$

4. Formula of Fick's law

$$dQ = DA\left(\frac{dc}{dx}\right)dt$$

Where, dQ is the mass flow (moles) during the time at across area A (sq. cm.), dc/dx the concentration gradient [moles/c.c. (cm)], and D, the proportionately constant or diffusion coefficient (sq. cm./sec). 'D' depends upon the property of the medium as well as he gas. It varies directly with the square of the absolute temperature and inversely with the total pressure.

5. Specific heat

We can calculate the specific heat of a soil C_s from the summation of the specific heat times the mass of the individual constituents;

Cs = $C_1 M_1 + C_2 M_2 + C_3 M_3 + C_4 M_4 + ...\ C_m M_n$ [cal/g(°C)]

Where,

Cs = Specific heat of soil

M_1, M_2, M_3, M_4 and $C_1, C_2 ......\ C_n$, mass and specific heat of individual constituents respectively.

6. Heat capacity

The heat capacity of as soil constituent is equal to its specific heat times, its density. The heat capacity of the soil per unit volume can be computed by following equation,

Cs = XsCs + XwCw + XaCa [cal/c.c.(°C)]

Where,

Cs = heat capacity of the soil

Xs, Xw and Xa are the volume fraction of the solid mass, water and air respectively.

Cs, Cw and Ca are the heat capacities of their respective constituents.

7. Thermal Conductivity and diffusivity

The charge of heat dQ/dt flowing into or out of the soil depends on the temperature gradient dT/dZ in the soil and the thermal conductivity Ks for any given soil volume the change of heat flow is given by,

$$dQ = Ks\left(\frac{dT}{dZ}\right)dt$$

Where,

Z is measured along the gradient.

K is thermal conductivity of soil expressed in J/°C/cm/sec.

The effect of a heat flow on the rate of change of temperature at a given depth depends on the difference between the heat flow into and out of a small volume of soil at that depth, and is given by,

$$\frac{dT}{dt} = \frac{k}{pc}\frac{d^2T}{dz^2}$$

Where,

p is the density of the soil,

c is the specific heat,

k/p^c is called thermal diffusivity of the soil and expressed in cm^2/sec.

8. Surface tension

Consider a water in a capillary tube having a boundary with air. The boundary layer between the water and the air is called meniscus. The meniscus is usually curved; it may make a definite angle – the angle of contact with the walls of the tube; and it put s the water column under a tension T, given by,

$$T = \frac{2\sigma}{\cos\alpha}$$

where,

r = radius of a circular tube

σ = surface tension of the water

α = angle of contact or angle of wetting

9. Formula of Soil Moisture by Gravimetric Method

$$\% \text{ moisture} = \frac{\text{weight of moist or wet soil} - \text{weight of oven dry soil}}{\text{weight of oven dry soil}} \times 100$$

10. Saturated flow

The flow of water under saturated conditions is determined by two major factors – the hydraulic force driving the water through the soil and the ease with which the soil pores permit water movement. This can be expressed as follows:

V = Kf

Where,

V = is the total volume of water moved per unit time

F = is the water moving force

k = is the hydraulic conductivity of the soil or permeability of the soil

11. Equation of hydraulic conductivity by Darcy

$$Qw = -\frac{(\Delta dw)\, mAt}{(\Delta ds)}$$

or
$$K = -\frac{Qw\,(\Delta dw)}{At\,(\Delta ds)}$$

where,

Qw = quantity of water in c.c.

K = rate constant or hydraulic conductivity in cm sec^{-1}

Δdw = water head in cm

A = soil area in sq. cm]

t = time in seconds

Δds = soil depth used in cm

12. Unsaturated flow

Presence of air phase in the soil system of unsaturated conditions the equation for the movement of water through the saturated soil can be modified as follows by introducing a dimensionless factor » with k. the factor λ varies from 0 to 1 and disappears when the soil is saturated. In this flow gravity, matric and pressure potentials are involved.

$V = k\,\lambda$ grad H.

13. Water use efficiency

The water use efficiency is defined mathematically as follows:

$$WUE = \frac{DW}{ET}$$

Where,

DW = Dry weight of crop per acre.

ET = water used in eavapotranspiration in acre inches per acre.

14. Formula of plastic flow law

The following equation shows that the volume flows I a function of the force applied.

$$V = K\mu \, (F - f)$$

Where,

V = Volume of flow

μ = coefficient of mobility

F = applied force

F = force necessary to overcome the cohesive forces of the system and just enough to start the flow (this force is termed "yield value")

K = Constant

15. Cohesive

The cohesive force of water film between two particles is given by the following equation:

$$F = \frac{k4\pi rT \cos\alpha}{D}$$

Where,

K = constant

r = radius of the particle

T = surface tension,

$\pm$ = distance between particles

F = cohesive force

16. Percent base saturation of soils

The percentage of the total cation exchange capacity (CEC) satisfied with basic cations is termed percent base saturation. It is defined as the extent to which the exchange complex of a soil is saturated with exchangeable cations other than hydrogen and aluminium and it is expressed as a percentage of the total cation exchange capacity.

$$\%BS = \frac{S}{T} \times 100$$

Where,

BS = Base saturation,

S = me of basic cations per 100 g soils

T = total exchange capacity me per 100 g soil

17. Ion-Exchange formulae

(1) Empirical formula

The Freundlich equation is one of the first used in soil studies and it states that,

$$m = AC^B$$

(2) Kinetic or statistical

The Langmuir equation is used to characterize ion adsorption particularly phosphorus in soils. The most common form of the Langmuir equation is,

$$m = \frac{ABC}{1 + B_C}$$

Where,

m = amount of ion adsorbed per unit weight of soil

c = ion concentration in soil solution to e considered

A = adsorption maximum

B = constant related to bonding energy

(3) Mass action

Gapon formula

All exchange equations are the same in the case of exchange of equal valence ion. This, however is not the case when the ions are of unequal valences like monovalent and divalent ions.

$$\frac{Nax/Ca}{Cax/Na^+} = K$$

Where,

x = is the exchange complex

K = is the constant

18. Soil formation equations

The transformation of a parent material (end product of weathering) to a soil material and to a soil profile is carried out by various factors and that were first put forwarded by Dockuchaiev (1889) in the form of equation:

S = f (cl, o, r, p, t......)

Where,

$\quad\quad$ S = soil formation

$\quad\quad$ f = function,

$\quad\quad$ cl = climate, $\quad\quad\quad\quad$ o = organisms,

$\quad\quad$ r = relief, $\quad\quad\quad\quad$ p = parent materials, $\quad\quad\quad$ t = time

Jenny then emphasized that a soil property is determined by the relative influence of all these factor. Jenny considered temperature and rainfall as climate; flora and fauna as biosphere organisms; elevation, slope/topography and depth of water table as relief. Jenny (1941) formulated the following equations:

$$S = f\ (cl,\ b,\ r,\ p,\ t......)$$

Where,

$\quad\quad$ B = biosphere (vegetation, organisms and man), and others are same as Dokuchaiev's equation.

In soil forming equation climate and biosphere are active factors, whereas, relief, parent material and time are passive factors.

19. Methods of expressing soil reaction

Soil reaction can be measured by determining the p^H. So p^H may be defined as the logarithm of the reciprocal of the H^+ ion actively and it can be expressed in moles per litre.

$$p^H = \log \frac{1}{AH} \text{ or } -\log AH^+$$

Where, AH^+ is the hydrogen ion activity in moles per litre.

20. Neutralizing value (N.V.) or Calcium Carbonate Equivalent (CCE)

Calcium carbonate equivalent (CCE) is defined as the acid neutralizing capacity of an agricultural liming material expressed as a weigh percentage of calcium carbonate.

$$\text{CCE of a liming material} = \frac{\text{Molecular weight of CaCO}_3}{\substack{\text{Molecular weight of a limingmaterial} \\ \text{whose CCE is to be determined}}} \times 100$$

21. *Per cent Effective Calcium Carbonate (ECC) or (Neutralising Index)*

The Effective Calcium Carbonate (ECC) rating of a limestone or liming materials is one product of its CCE and the fineness factor. The fineness factor is the sum of the product of the percentage of material in each of the three size fractions multiplied by the appropriate effectiveness factor.

Per cent ECC or N.I. = CCE x fineness factor

22. *Sodium Adsorption Ratio (SAR)*

The U.S. salinity laboratory developed the concept of Sodium Adsorption Ratio (SAR) to define the equilibrium between soluble and exchangeable cations as follows:

$$SAR = \frac{Na^+}{\sqrt{\dfrac{Ca^{2+} + Mg^{2+}}{2}}}$$

(Na^+, Ca^{2+} and Mg^{2++} are concentrations in saturation extract in mel^{-1})

the value of SAR can be used for the determination of Exchangeable sodium percentage (ESP) of the saturation extract by using the following formula:

$$ESP = \frac{100\,(-0.0126 + 0.01475\,SAR)}{1 + (-0.0126 + 0.01475\,SAR)}$$

Sometimes the following regression equation is used for the appraisal of alkali soil by determining the value of ESP from the value of SAR.

Y = 0.0673 + 0.035X

Where,

Y = indicates ESP and

X = indicates SAR

Soils having SAR value greater than 13 are considered as alkali or sodic soil.

23. *Leaching Requirement (LR)*

Leaching Requirement can be calculated by formula

$$LR = \frac{Ddw \times 100}{Diw\ ECiw} = \frac{ECiw \times 100}{ECdw} \text{(LR expressed in percentage)}$$

Where,

Ddw = Depth of drainage water in inches

Diw = Depth of irrigation water in inches

ECiw = Electrical conductivity of the irrigation water in dsm^{-1}

ECdw = Electrical conductivity of the drainage water in dsm^{-1}

24. Gypsum Requirement (GR)

Gypsum Requirement is determined from the formula,

$$\text{Gypsum Requirement (GR)} = \frac{[\text{ESP (initial)} - \text{ESP (final)}] \times \text{CEC}}{100}$$

i.e. me of $Ca^{2+}/100$ g soil

ESP (final) is obtained from the analysis of soil before reclamation or application of gypsum; ESP (final) is usually kept at 10 and CEC is the cation exchange capacity in me/100g or C mol (p$^+$) kg^{-1} of the soil.

25. Residual Sodium Carbonate (RSC)

Residual Sodium Carbonate is used to evaluate the quality of irrigation water and is expressed in mel^{-1}.

RSC (mel^{-1}) $(CO_3^{2-} + HCO_3^{-}) - (Ca^{2+} + Mg^{2+})$

26. Chloride concentration

$$\text{chloride (cl}^-\text{) concentration (mel}^{-1}) = \frac{Cl^-}{-CO_3^{2-} + HCO_3^- + SO_4^{2+} + Cl^- + NO_3^-}$$

27. Soluble Sodium Percentage (SSP)

$$SSP = \frac{Na \times 100}{Ca + Mg + Na}$$

28. Magnesium hazard

$$Mg - adsorption = \frac{Mg^{2+}}{Ca^{2+} + Mg^{2+}}$$

29. Growth expression equations

The plant is a product of both its genetic constitution (potential for maximum growth) and its environment (variable). Therefore, growth may be expressed as follows :

$$G = f(x_1, x_2, x_3, x_4 \ldots\ldots, x_n)$$

Where,

G = some measure of plant growth and $x_1, x_2, x_3, x_4 \ldots\ldots, x_n$ the various growth factors. Again, if all but one of the growth factor is present in adequate amounts, an increase in the quantity of this limiting factor will generally result in increasing plant growth as follows:

$$G = f(x_1) \, x_2, x_3, x_4 \ldots\ldots, x_n$$

However, this is not a simple linear relationship.

30. Mitscherlich's growth equation

Mathematically Mitscherlich's growth equation can be expressed as follows:

$$\frac{dy}{dx} = (A - y)C$$

where,

dy is the increase in yield resulting from an increment,

dx of the growth facto x,

A is the maximum possible yield resulted from the optimum supply of all growth factors.

y is the yield obtained after any given quantity of the factor x has been applied and c is the proportionately constant (considered as efficiency coefficient or factor).

31. Spillman's equation

W.J. Spillman expressed the relation as follows:

$$y = M(1 - R^x)$$

where,

y is the amount of growth produced by a given quantity of the growth factor x,

x is the quantity of the growth factor,

M is the maximum possible yield, and

R is a constant.

By combining it with Mitscherlich equation, Spillman developed a reduced form of equation as follows:

$$y = A(1 - 10^{-cx})$$

$$y = A - A \cdot 10^{-cx}$$

or $\quad A - y = A \cdot 10^{-cx}$

or $\quad \log (A - y) = \log A + (-cx) \log 10$ (taking log of both sides)

or $\quad \log (A - y) = \log A - cx \, 1$

or $\quad \log (A - y) = \log A - c \, (x)$

Where,

y is the yield produced by a given quantity of the growth fact x,

A is the maximum possible yield, and

c is a constant which depends on the nature of the growth factor

32. Bray's nutrient mobility equation

Bray has modified the Mitscherlich equation to:

$$\log (A - y) = \log A - C_1 b - C_x$$

Where, A, y and x are maximu yield, yield obtained and amount of added fertilizer nutrient respectively.

b $\;=$ amount of an imm0obile but available form of nutrients (like p and k)

$C_1 =$ constant or efficiency factor of b for yields,

$C \;=$ constant or efficiency factor of x

33. Activity Index (AI)

An activity index is used by the Association of official Agricultural chemists to evaluate the suitability of urea formaldehyde compounds as follows:

$$\text{Activity Index (AI)} = \frac{\%CWIN - \% HWIN}{\% CWIN} \times 100$$

Where,

CWIN = percent nitrogen insoluble in cold water (25°C)

HWIN = percent nitrogen insoluble in cold water (98-100°C)

34. Aeration status of a submerged soil

Immediately after submergence, the normal process of gaseous exchange between soil and air is restricted. The magnitude of this effect is evident from the relative values in air and in water of the diffusion coefficient D, in the following equation,

$$V = aD \, (T/T_O)^2 \cdot dp/dl$$

Where,

$$V = \text{Volume of gas, c.c. } cm^2 \ sec^{-1}$$

$$a = \text{porosity factor}$$

$$dp/dl = \text{pressure gradient}$$

$$T = \text{absolute temperature}$$

35. Bulk density of saturated soil

When soil is compacted to such a degree that all voids (pore spaces) are filled with water, no air voids (spaces) are present, and no soil water is expelled from the voids, the soil is saturated and its bulk density is maximum. So the bulk density of saturated soil can be calculated from the following equation:

$$PB = \frac{P_p rw}{L + \dfrac{P_p rw}{100}}$$

Where,

$$PB = \text{bulk density,}$$

$$P_p = \text{particle density,}$$

$$W = \text{water content and}$$

$$rw = \text{unit weight of water}$$

(36) $$\%P = P_2O_5 \times 0.43$$

$$\%P_2O_5 = \% \ P \times 2.29$$

(37) $$\%K = K_2O \times 0.83$$

$$\%K_2O = \%K \times 1.2$$

The relationship between mass number (m), atomic number (a) and the neutron number (n) can be written as,

$$m = a + n$$

As protons and neutrons are the main constituents of the nucleus, they are combinedly known s nucleons.

36. Max Planck in 1900 proposed that a black body radiates energy not continuously but discontinuously in discrete energy packets, called quanta, given by the relation as follows,

$$\varepsilon = h\nu = \frac{hc}{r}$$

where,

ε = quantum of energy in joules

v = frequency of the emitted radiation in cycles per sec

c = velocity of light (3.0×10^8 m/sec.)

h = Planck's constant (6.63×10^{-34} Joules sec)

r = wavelength of the electromagnetic radiation emitted in meters

37. Forces in the nucleus

The nuclear binding energy, Eb/A, per nucleon for element of mass number, greater than 20 is approximately constant, about 8 Mev (million electron volt), i.e., Eb = a constant x A (Atomic mass). The total nuclear binding energy given by the following expression:

$$E_b = 14.1\ A - 13\ A^{2/3} - 0.6Z^2/A^{1/3}$$

Where,

Z = atomic number

[1 Mev = 2.31×10^{10} cal per mole]

38. Celendenin's Equation

The following Celendenin's equation can be used for calculating the range of beta particles.

$$R_{max}\ (mg\ cm^{-2}) = 542\ E_{max} - 133\ (MeV)$$

Where,

R_{max} = maximum range of beta particles,

E_{max} = maximum energy,

Above equation can be applicable where E_{max} value is 0.8 MeV.

39. Half-life period

The time required for a given amount of a radio element to decay to one-half its initial value is called its half-life. Hence, the half-life period T is given by,

$$T = \frac{2.030}{K}\log^2$$

$$T = \frac{0.693}{K}$$

Where, K is disintegration/decay constant.

40. E-value

E-value concept was developed by Russell. This method is a direct application of the isotopic dilution principle. The amount of nutrient in the soil at equilibrium with the same nutrient in the soil solution can be measured with E-value. Reaction can be depicted as follows

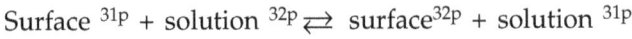

Surface ^{31}p + solution ^{32}p $\rightleftarrows$ surface^{32}p + solution ^{31}p

At equilibrium,

$$\frac{surface^{31P}}{solution^{32P}} = \frac{surface^{32P}}{solution^{32P}}$$

41. A-value

A-value concept was developed by Fried and Dean (1952) when the plant is confronted with two sources of a given nutrient, the plant will absorb from each of these in proportion to the respective amounts available. The amount of available nutrient in soil to be determined in terms of fertilizer standard is known as 'A' value. Mathematically A value can be expressed as:

$$A = \frac{B(1-Y)}{Y}$$

Where,

A = amount of available nutrient in soil

B = Amount of applied fertilizer nutrient

Y = proportion of nutrient in the plant derived from the fertilizer nutrient

A-value concept is expressed as follows:

'A' value = [% pdfs/%pdff] × rate of P application (kg/ha)

Where,

%pdff = % of total p in the plant derived from fertilizer

%pdfs = (100 – %pdff)

42. L-value

L-value was first suggested by Larsen (1952). If labeled fertilizer phosphorus is added to a soil at different levels or rates and plants are grown, the specific activity analysis of the plant material gives a constant value which is attributable to the equilibrium between the applied phosphate and exchangeable phosphate in the soil. The amount so determined is known as L-value and also termed as "Labile form" of phosphorus. Larsen (1952) used the following equation for isotopic dilution to calculate isotopically labile p or L-value.

L-value = [(ao/ap) − 1] × Applied fertilizer p

Where, ao and ap are specific activities of the applied p and phosphorus in plant respectively.

46. *The major factors affecting water erosion are climate, topography, vegetation and soils*

This can be written as,

Ew = f (c, t, v, s)

Where, Ew = Erosion due to action of water

c = climate

t = topography

v = vegetation and

s = soils

44. *Equivalent diameter*

Erodibility of soil can be related with the equivalent diameter of the particles. The equivalent diameter is equal to the product of bulk density and the diameter of soil particles divided by the particle density (2.65) of soil.

$$\text{Equivalent diameter (de)} = \frac{\text{Bulk density} \times \text{diameter of the soil particle}}{\text{Particle density (2.65)}}$$

The most erodible soil particles are about 0.1 mm in equivalent diameter.

45. *Diffusion coefficient*

It is one of the most important factors that control the movement of ions from the soil to the root. It can be expressed as follows:

$$De = Dw\, f\theta\frac{1}{B}$$

Where,

De = Effective diffusion coefficient

Dw = Diffusion coefficient of ions in water

θ = volumetric water percentage

f = Tortuosity factor (zig-zag path in the soil system)

b = Buffering capacity of soil

46. ESP (Exchangeable Sodium Percentage)

The percentage of the cation exchange capacity of a soil occupied by sodium is known as exchangeable sodium percentage and is expressed as follows:

$$ESP = \frac{\text{Exchangeable sodium (me/100 g)}}{\text{CEC (me/100 g)}} \times 100$$

47. Universal Soil Loss Equation (USLE)

It is defined as the loss of soil from the product of following six factors and expressed as:

$$A = R.K.L.S.C.P$$

Where,

A = Erosion soil loss in t acre^{-1} year^{-1}

R = Rainfall factor

K = Soil erodibility factor

L = Slope length factor

S = slop gradient factor (percent steepness)

C = vegetative cover and management factor

P = practices used for erosion control

48. To get target yield of crop then the crop wise fertilizer prescription equations can be used as given below

Following abbreviations used in the following equations

FN = Fertilizer nitrogen to be added

FP_2O_5 = Fertilizer phosphorus to be added

FK_2O = Fertilizer potassium to be added

T = Target yield

SN = Soil test nitrogen value

SP = Soil test phosphorus value

SK = Soil test potassium value

FYM = Farm yard manure to be added

(1) Jawar (Irrigated)

$$FN = 4.04T - 0.22\ SN$$

$$FP_2O_5 = 2.72\ T - 8.26\ SP$$

$$FK_2O = 3.82\ T - 0.17\ SK$$

(2) Jawar (Dry land)

$$FN = 4.58T - 0.96 \ SN$$
$$FP_2O_5 = 2.21 \ T - 6.94 \ SP$$
$$FK_2O = 3.34 \ T - 0.22 \ SK$$

(3) *Rabi* Jawar

$$FN = 4.70 \ T - 0.77 \ SN$$
$$FP_2O_5 = 2.00 \ T - 4.29 \ SP$$
$$FK_2O = 3.35 \ T - 0.33 \ SK$$

(4) Bajara

$$FN = 3.31 \ T - 0.38 \ SN$$
$$FP_2O_5 = 3.38 \ T - 4.41 \ SP$$
$$FK_2O = 1.65 \ T - 0.06 \ SK$$

(5) Groundnut

$$FN = 4.16 \ T - 0.37 \ SN$$
$$FP_2O_5 = 4.96 \ T - 4.36 \ SP$$
$$FK_2O = 3.14 \ T - 0.16 \ SK$$

(6) Summer Groundnut

$$FN = 1.18 \ T - 0.40 \ SN$$
$$FP_2O_5 = 8.23 \ T - 6.15 \ SP$$
$$FK_2O = 3.22 \ T - 0.10 \ SK$$

(7) Tur

$$FN = 5.61 \ T - 0.54 \ SN$$
$$FP_2O_5 = 5.72 \ T - 4.73 \ SP$$
$$FK_2O = 6.33 \ T - 0.17 \ SK$$

(8) Mung

$$FN = 4.56 \ T - 0.18 \ SN$$
$$FP_2O_5 = 12.51 \ T - 7.61 \ SP$$
$$FK_2O = 3.53 \ T - 0.05 \ SK$$

(9) Sunflower

$$FN = 3.94 \ T - 0.61 \ SN$$
$$FP_2O_5 = 7.18 \ T - 6.82 \ SP$$
$$FK_2O = 4.82 \ T - 0.12 \ SK$$

(10) Gram

$$FN = 5.25 \ T - 0.46 \ SN$$

$$FP_2O_5 = 3.87 \ T - 2.77 \ SP$$

$$FK_2O = 1.29 \ T - 0.04 \ SK$$

(11) Wheat

$$FN = 7.54 \ T - 0.74 \ SN$$

$$FP_2O_5 = 1.90 \ T - 2.88 \ SP$$

$$FK_2O = 2.49 \ T - 0.22 \ SK$$

(12) Cotton

$$FN = 13.1 \ T - 0.75 \ SN$$

$$FP_2O_5 = 6.83 \ T - 2.84 \ SP$$

$$FK_2O = 8.57 \ T - 0.14 \ SK$$

(13) Sugarcane (Suru)

$$FN = 4.39 \ T - 1.56 \ SN$$

$$FP_2O_5 = 1.62 \ T - 4.56 \ SP$$

$$FK_2O = 1.86 \ T - 0.37 \ SK$$

(14) Sugarcane (Pre-seasonal)

$$FN = 5.40 \ T - 1.08 \ SN$$

$$FP_2O_5 = 2.60 \ T - 6.51 \ SP$$

$$FK_2O = 1.90 \ T - 0.15 \ SK$$

(15) Sugarcane (Adsali)

$$FN = 4.76 \ T - 1.34 \ SN$$

$$FP_2O_5 = 1.24 \ T - 1.55 \ SP$$

$$FK_2O = 2.73 \ T - 0.21 \ SK$$

(16) Onion *(Rabi)*

$$FN = 5.40 \ T - 0.54 \ SN$$

$$FP_2O_5 = 4.00 \ T - 4.32 \ SP$$

$$FK_2O = 3.10 \ T - 0.13 \ SK$$

(17) Soyabean

$$FN = 4.30 \ T - 0.38 \ SN$$

$$FP_2O_5 = 9.76 \ T - 8.17 \ SP$$

$$FK_2O = 2.80 \ T - 0.06 \ SK$$

(18) Rice (upland)

$$FN = 5.52 \ T - 0.54 \ SN$$

$$FP_2O_5 = 2.19 \ T - 0.83 \ SP$$

$$FK_2O = 2.37 \ T - 0.05 \ SK$$

(19) Rice (lowland)

$$FN = 5.20 \ T - 0.34 \ SN$$

$$FP_2O_5 = 9.40 \ T - 13.66 \ SP$$

$$FK_2O = 2.73 \ T - 0.16 \ SK$$

(20) Brinjal (Krishna) without FYM

$$FN = 4.82 \ T - 0.53 \ SN$$

$$FP_2O_5 = 3.14 \ T - 7.32 \ SP$$

$$FK_2O = 3.21 \ T - 0.05 \ SK$$

(21) Brinjal (other varieties) with FYM

$$FN = 4.27 \ T - 0.47 \ SN - 2.93 \ FYM$$

$$FP_2O_5 = 2.85 \ T - 2.87 \ SP - 1.19 \ FYM$$

$$FK_2O = 2.86 \ T - 0.09 \ SK - 2.09 \ FYM$$

(22) Cabbage (*Rabi*) (Golden acre)

$$FN = 8.28 \ T - 0.21 \ SN$$

$$FP_2O_5 = 4.72 \ T - 2.34 \ SP$$

$$FK_2O = 6.68 \ T - 0.19 \ SK$$

(23) Chilli (*Rabi*) (Phule Jyoti) Without FYM

$$FN = 50.23 \ T - 0.54 \ SN$$

$$FP_2O_5 = 27.09 \ T - 3.17 \ SP$$

$$FK_2O = 36.48 \ T - 0.30 \ SK$$

(24) Chilli (*Rabi*) (Phule Jyoti) with FYM

$$FN = 37.25 \ T - 0.40 \ SN - 3.38 \ FYM$$

$$FP_2O_5 = 25.40 \ T - 2.97 \ SP - 1.88 \ FYM$$

$$FK_2O = 34.00 \ T - 0.26 \ SK - 1.66 \ FYM$$

(25) Okra (Summer) Without FYM

$$FN = 16.86 \ T - 0.45 \ SN$$

$$FP_2O_5 = 10.31 \ T - 2.36 \ SP$$

$$FK_2O = 11.60 \ T - 0.15 \ SK$$

(26) Okra (Arka Anamika) with FYM

$$FN = 15.54 \ T - 0.42 \ SN - 2.32 \ FYM$$

$$FP_2O_5 = 9.61 \ T - 2.21 \ SP - 1.45 \ FYM$$

$$FK_2O = 11.06 \ T - 0.14 \ SK - 1.46 \ FYM$$

(27) Turmeric *(Kharif)* without FYM

$$FN = 11.10 \ T - 1.78 \ SN$$

$$FP_2O_5 = 4.54 \ T - 7.55 \ SP$$

$$FK_2O = 5.40 \ T - 0.545 \ SK$$

(28) Turmeric (Var. Salem) with FYM

$$FN = 6.45 \ T - 0.88 \ SN - 2.55 \ FYM$$

$$FP_2O_5 = 4.03 \ T - 6.48 \ SP - 0.59 \ FYM$$

$$FK_2O = 4.52 \ T - 0.45 \ SK - 1.40 \ FYM$$

(29) Soybean *(Kharif)* without FYM

$$FN = 6.86 \ T - 0.68 \ SN$$

$$FP_2O_5 = 6.17 \ T - 4.46 \ SP$$

$$FK_2O = 3.96 \ T - 0.13 \ SK$$

(30) Soybean (var. JS-335) with FYM

$$FN = 3.97 \ T - 0.39 \ SN - 0.09 \ FYM$$

$$FP_2O_5 = 4.14 \ T - 2.95 \ SP - 1.50 \ FYM$$

$$FK_2O = 3.47 \ T - 0.11 \ SK - 0.27 \ FY$$

(31) Cauliflower *(Rabi)* without FYM

$$FN = 6.83 \ T - 0.35 \ SN$$

$$FP_2O_5 = 4.25 \ T - 2.21 \ SP$$

$$FK_2O = 3.90 \ T - 0.08 \ SK$$

(32) Cauliflower (Var. Namdhari No. 90) with FYM

$$FN = 6.0 \ T - 0.30 \ SN - 1.44 \ FYM$$

$$FP_2O_5 = 3.92 \ T - 2.04 \ SP - 1.20 \ FYM$$

$$FK_2O = 3.07 \ T - 0.06 \ SK - 1.12 \ FYM$$

(33) Tomato (Summer) without FYM

$$FN = 5.33 \ T - 0.46 \ SN$$

$$FP_2O_5 = 3.88 \ T - 4.16 \ SP$$

$$FK_2O = 5.16 \ T - 0.25 \ SK$$

(34) Tomato (Var. Dhanshree) with FYM

$$FN = 4.13\ T - 0.43\ SN - 1.13\ FYM$$

$$FP_2O_5 = 2.50\ T - 2.78\ SP - 0.57\ FYM$$

$$FK_2O = 3.44\ T - 0.22\ SK - 75\ FYM$$

(35) Potato *(Rabi)*

$$FN = 1.549\ T - 0.40\ SN$$

$$FP_2O_5 = 0.906\ T - 5.53\ SP$$

$$FK_2O = 1.315\ T - 0.17\ SK$$

(36) Potato (Var. Khufri Jyoti)

$$FN = 1.2073\ T - 0.315\ SN - 0.81\ FYM$$

$$FP_2O_5 = 0.878\ T - 5.35\ SP - 0.71\ FYM$$

$$FK_2O = 1.180\ T - 1.156\ SK - 0.76\ FYM$$

REFERENCES

Smith and Cristol, 1971. *Organic Chemistry.* Affiliated East-West Press Pvt. Ltd. New Delhi.

Sehgal, J. 2000. *Pedology. Concepts and Applications.* Kalayani Publishing Co., New Delhi.

Yawalkar, K.S., Agarwal, J. P. and Bokde, S. 2002. *Manures and Fertilizers.* Agri. Horticultural Publishing House, Nagpur.

Havlin, J.L., Beaton, J. D., Tisdale, S. L. and Nelson, W. L. 2005. *Soil Fertility and Fertilizers.* Dorling Kindersley, Pvt. Ltd. New Delhi.

Singh, D., Chhonkar, P. K. and Dwivedi, B. S. 2007. *Manual on Soil, Plant and Water Analysis.* Westville Publishing House. New Delhi.

Mee, A.J. 1969. *Physical Chemistry.* Heinemann Educational Books Pvt. Ltd. London.

Das, D. 1996. *Introductory Soil Science.* Kalayani Publishing Co., New Delhi.

Dagi, D.A., Kadam, J.R. and Patil, N. D. 1996. *Textbook of Soil Science.* Media Promoters and Publisher Pvt. Ltd. Bombay.

Ghildyal, B.P. and Tripathi, R.P. 2001. *Soil Physics.* K.K. Gupta for New Age International (P) Ltd. Publisher, New Delhi.

Daniel Hillel. 2004. *Introduction to Soil Physics.* Published by Elsevier a division of Reed Elsevier India Private Ltd., New Delhi.

Brady, N.C. 2004. *Nature and Properties of Soil.* Pearson Education Singapore.

Abalaka, J.A. 1999. Assuring food quality and safety : Back to the basis quality control throughout the food chain. FAO/WHO/WTO Conf. On Int. Food Trade beyond 2000. Melbourne, Australia, II-IS Oct., 1999.

Abalaka, J.A. 1999. *Assuring food quality and safety: Back to the basis quality control throughout the food chain.* FAO/WHO/WTO Conf. On Int. Food Trade beyond 2000. Melbourne, Australia, 11-15 Oct., 1999.

Acharya, C.N., 1939. *The hot fermentation process for composting town refuse and other waste materials.* Indian J. agric. Science. 9:741-817.

Agarkar, O.D. 2002. *Use of Azotobactor for sustainable crop production winter school, INM,* Dept. of Agri. Chem. & Soil Sci., Dr. Panjabrao Deshmukh Krishi vidyapeeth Akola P.P. 146-155

Agriculture Canada. 1989. *Research in Sustainable Agriculture.* Proceedings of a workshop 3-4 February, 1989 at Saint-Hyacinthe, QC.

Agriculture Canada. 1989a. *Growing Together: A Vision for Canada's Agrifood Industry.* Supply & Services Canada, Ottawa. Pp. 41.

AIFOF. 1996. *Soil management.* All India Federation of organic farming, Thane (India) p.p. 10-12

Alagwadi, A.R. and Kulkarni, J.H. 1993. *Biofertilizers An integrated approach for sustainable agriculture.* Nal. Conf. Biofertilizers and organic farming, Ministry of Agriculture, Govt. of India and Dept. of Agri. Govt. of Tamilnadu Madras p.p. 1-4.

Amar Chandra, 1998. *Research and development in vegetable production through organic farming.* Notes on training programme on organic agriculture. An ultimate food production system held at JNKVV, Jabalpur, Sept. 8-22, 1998.

Amar Chandra. 1998. *Research and development in vegetable production through organic farming.* Notes on training programme on organic agriculture. An ultimate food production system held at JNKVV, Jabalpur, Sept. 8-22, 1998.

Amrutsagar, V.M. and P.B. Shinde. 2004. *Organic farming for sustainability Rahuri.* Chapter ISSS, Souvenir, PP 78-88.

Anthonis, G. 1994. *Agro Chem.* News in Brief 17(2): 12-15.

Arner, R. 1995. *George Washington: the composer of the country.* Washington post. Sunday Free for All Sept. 10, 1995, A-20.

Asalmol, M.N. 2002. *Use of VAM for sustainable crop production.* Winter Schchool, INM. Dept. of Agri. Chem. And Soil Sci. Dr. Panjabrao Deshmukh Krishi Vidyapeeth Akola. p.p. 212-214.

Baghel, B.S., Sarnaik, D.A. and Pathak, A.C. 1986. *Intercropping of legumes in basrai banana.* Research and Development Report, 3(2): 7-9.

Baphana, P. D. 1992. *Organic Farming in Sapota.* Proc. National Seminar Organic farming, Mahatma Phule, Krishi Vidyapeeth, Pune. p.p. 55-57.

Basavannepppa M.A. and Biradar D.P. 2003. Kisan world, 61-62.

Bergstrom Lars and Kirchmann, Holger. 2002. 17th World Congress of Soil Science Bangkok, Thai-land Abs. Vol V p.1707

Beyl, C.A. 1992. *Rachel Carson, Silent spring and the environmental movement.* Proceedings of the workshop on History of the organic movement held at 88th ASHS Annual meeting. The Pennsylvania state University, University Park. U.S.A. on 24-July- I 991.

Bhale, R.I. 1992. *Vermiculture Biotechnology Basis, scope for Application and Development.* Proc. Natn. Sem. Organic Farming, .Mahatma Phule Krishi Vidyapeeth, Pune, p.p. 50-51.

Bhardwaj, K.K.R. and Gaur, A.C. 1985. *Receding of organic wastes.* ICAR Publ. New Delhi. Pp. 5.

Bhardwaj, K. K. R. and Gaur, A. C. 1985. *Recycling of Or-ganic Wastes.* ICAR Publication, New Delhi p.5.

Borlaug, N. 2002. *CNS News, Com.* http:\\www.sclentific all iance.com\gm_organic_food.

Borlaug, N. 2002. *CNS News, Com.* http:\\www.scientific all iance.com\gm_organic_food.

Borlaug, Norman. 2002. CNS News Com May 01: 2002 (http://www.scientificalliance.com/news/gm_organic_food/organic_forests.030502.html).

Borwne, C.A. 1942. *A source book of agricultural chemistry Chronica Bitanica* 8(1): 1-290.

Buyanovsky, G.A. and G.H. Wagner. 1998. Bio, Fert. Soils, 27 ; 242.

Carter, V.G. and Dale, T. 1974. Topsoil and Civilization. University of Oklahoma Press, Norman, Oklahoma.

Charter, R.A., Tabatabai, M.A. and Schafer W. 1993. Metal contents of fertilizers marketed in Lowa. Communications in Soil Science and Plant Analysis 24, 961-972.

Chen, Y, Chaung, M. and Aus. S. 1997. Effect of VAM fungi on growth promotion of citrus. J. Agric. Res., China, 46:324-332.

Chen, Y, Chaung. M. and Aus. S. 1997. Effect of VAM fungi on growth promotion of citrus. 1. *Agric. Res., China, 46:324-332.*

Chhonkar, P.K. (1995) In Soil Nutrient Management (P.K. Chhonkar, G. Narayanasamy and R.K. Rattan Eds.) Ministry of Agriculture, Govt. of India and Indian Agricultural Research Institute, New Delhi.

Chhonkar, P.K. 1994a. In: Phosphorus Research in India (G. Dev Ed.) Potash sand Phosphate Instt. of Canada: India Programme. pp 120- 125.

Chhonkar, P.K. 1998. Biological strategies for nutrient cy-cling in soil system In: Soil Plant Microbe Interac-tion in relation to Integrated Nutrient Management (B. D. Kaushik Ed.) Indian Agril. Research Institute, New Delhi pp. 48-54.

Chhonkar, P.K. 2001. Improving phosphorus utilization efficiency through microbial intervention In: Phos-phorus in Indian Agriculture: Issues and Strate-gies (K.N. Tiwari, R.K. Rattan, P.K. Chhonkar Eds.) Potash and Phosphorus Institute of Canada India

Chhonkar, P.K. 2002. Organic farming - myth and reality. In: FAI Annual .Seminar, Fertilizers and Agricul-ture: Meeting the challenges. Fertilizer Association of India, New Delhi. pp S III-3-9.

Chhonkar, P.K. 2002. Soil research in India: Some oversights and failures. 1. *Indian Soc. Soil Sci.* 50(4):328-332.

Chhonkar, P.K. 2002. Soil research in India: Some oversights and failures. J. Indian Soc. Soil Sci., 50(4):328-332.

Chhonkar, P.K. and Tilak, K.V.B.R. 1997. Biofertilizers for sustainable agriculture: Research gaps and future needs. In: Plant Nutrient Needs, Supply, Efficiency and Policy Issues: 2000-2025 (J.S. Kanwar and J.E. Katyal Eds.) National Academy of Agricultural Sciences pp. 52-66.

Chhonkar, P.K. Datta, S.P., Joshi, H.E. and Pathak, H. 20001. Impact of industrial effluents on soi I health and agriculturel. Distillery and paper mill effluent. Journal of Scientific and Industrial Research 59, 350-361

Chhonkar, P.K. l994b. Crop response to phosphatic bio-f-ertilizers. Fertilizer News 39, 41.

Chhonkar, P.K., Datta, S. P. Joshi H.C. and Pathak, H. 2000b. Impact of industrial effluents on soi I health and agriculture II. Tennery and textile effluents. Journal of Scientific and Industrial Research 59: 446-454.

Chhonkar, R.K. 2001. The state of Indian Soil Science and Challenges to be faced during Twenty- First Century. J. Indian Soc. Soil Sci. 4a (4); 532-536

Chitdeshwari, T. and Pravikesavan. 1998. Effect of organic and inorganic nitrogen sources and their combination on crop yield and soil fertility. In : Integrated plant supply system for sustainable productivity (Eds : Acharya, C. N. *et al.),* Indian Institute of Soil Science, Bhopal, pp 147-151.

Chitdeshwari. T. and Pravikesavan. 1998. Effect of organic and inorganic nitrogen sources and their combination on crop yield and soil fertility. In: Integrated plant supply system for sustainable productivity (Eds, Acharya. C.N. *et al.).* Indian Institute of Soil Science, Bhopal, pp 147-151.

Christianson, R. 1988. A marketing plan for a sustainable food system. In: J. Hartman (Editor) An organic food system for Canada : proceedings of a conference. Canadian organic Growers, Ottawa Ont. pp:7-17.

Clark. F.E. 1957. Living organisms in the soil In: US Dept. of Agriculture Year Book 1957 soil (A. Steffered Ed.) pp 157-164.

Combs, S.M., Peters, J.B. and Jhang J.S. 1998. Manure: Trace element content and impact on soil manage-ment. 15th Soil / plant analysis Workshop (NCR-13). Council on Soil Testing and Plant Analysis, Novem-ber 10-1 I St. Louis No, USA.

Conford, P. (Ed.). 1988. The organic tradition, an anthology of writings on organic farming 1900-1950. Green Books, Devan, England.

Consultation Organisation, New Delhi pp.148+viii.

Copperband, L.R. 2000. Composting; Art and Science of organic waste conversion to a valuable soil resource. Laboratory Medicine. 31 :283-290

Dahama, A.K. 1996. Agricultural waste management and crop production in book. Organic farming for sustainable Agriculture Published by Agro Bot Publishers, Bikaner pp 131-157

Dhyan Singh. 2003. Personal communication. Fertilizer Statistics 2000-01, Fertilizer Association of In-dia. New Delhi

Dixit, A.I., Patil, V.H. and Singh, O.P. 1992. Role of Biofertilizers in rice production in Konkan region of Maharashtra Proc. Nal. Sem. Organic farming, Mahatma Phule Krishi Vidyapeeth, College Of Agriculture, Pune p.p. 98-99.

Dixit, P.K. 1993. Relevance of organic Farming in Modern Era; Natn. Conf. Biofertilizcrs and organic farming Department of Agriculture Cooperation, Govt. of India and Department of Agriculture, Government of Tamilnadu, Chennai, p.p. 93-95

Earnest, R. and Buffingston, L.E. 1976. Crop residues CRC Hand Book of Biosolar Resources vol. II.

Eberhardt, D.L. and Pipes, W. O.1972. Iions composting applications. In large scale composting. Edited by M. J. Satriana, Noyes Data Corporation, Park vidge, New Jersey.

Edwards, C.A. 1987. The concept of integrated systems in lower input/sustainable agriculture. American J. Alternative Agriculture 2: 148-152.

Eswaran, H. and Virmani, S. M. 1990. The soil component in sustainable Agriculture. Sustainable Agriculture: Issues perspectives and prospects in Semi Arid Tropics (2): 64-86.

Foundation, Cochin, Kerala 354 p.

Fukuoka, M. 1985. The Natural Way of Farming. Japan Publications. Tokyo and London. 280 pp.

Funtilana S. 1990. Safe, inexpensive, profitable and sen-sible. International Agricultural Development, March-April 24.

Gaur, A.C., Neelakantan, S. and Dargan, K.S. 1984. Organic manures. ICAR, New Delhi. pp. 159.

Gaur, A.C., Sadasivan, K.V., Mathur, R.S. and Magu, S. P. 1982. Rote of mesophgllic fungi in composting Agricultural wastes 4: 453-460.

Ghonsikar, C.P. and Shinde, V.S. 1997. Nutrient Management practices in pulses and pulse based cropping system(Ed), Scientific Publishers (India) p.p. 91-124.

Goswami, N.N. 1998. Some thoughts on the concept, relevance and feasibility of IPNS under Indian conditions. In: Integrated plant nutrient supply system for sustainable productivity (Eds. Acharya C.N. *et al.*). Indian Institute of Soil Science. Bhopal. pp 3-9.

'Goswami, N.N. 1998. Some thoughts on the concept, relevance and feasibility of IPNS under Indian conditions. In : Integrated plant nutrient supply system for sustainable productivity (Eds. Acharya C.N. *et al.*). Indian Institute of Soil Science, Bhopal, pp 3-9.

Gupta, S.B. 1995. Effective utilization of phosphorus in rice-wheat cropping system in a Vertisol through VA-mycorrhizae and phosphorus solublizer. Ph.D. Thesis, Post Graduate School. Indian Agricultural Research Institute, New Delhi. Pp.131 + XXII.

Hall, A.D. The Book of the Rothalllsted Experiments (2d ed. 1917); Sir Edward John Russell, British Agricultural Research: Rothamsted (ed. 1947); catalogue of the station's publications, ed. by D. H. Boalch (1954).

Hangarge, D.S., Raut, R.S., Malewar, G.U., More, S.D. and Keshbhat, S.S. 2002. Yield attributes and nutrient uptake by chilli due to organic and inorganic on Vertisol. J. Maharashtra agric. Univ. 27(1):109.

Hangarge, D.S., Raut, R.S., Malewar, G.O., More, S.D. and Keshbhat, S.S. 2002. Yield attributes and nutrient uptake by chilli due to organic and inorganic on Vertisol. 1. *Maharashtra agric. Vniv.* 27(1): 109.

Hardter, R. and Krauss, A. 1999. Balanced fertilization and crop quality. IFA Agric. Conf. On managing plant nutrition. Barcelona, Spain. June 29 July 2. 1999.

Hardter, R. and Krauss, A. 1999. Balanced fertilization and crop quality. IFA Agric. Conf. On managing plant nutrition. Barcelona, Spain. June 29 July 2. 1999.

Harwood, R.R. 1984. Organic farming Research at the Rodale Research Centre In: Proceeding of a symposium on organic farming : current Technology and Its role in a sustainable agriculture, ASA special publication No. 46. pp: 1-17.

Hegde, R., Korikanthinath, V.S. and Mulge, R. 1995. Scope for organic farming in plantation crops. In-dian Coffee 59, 11-12.

Hicks C.S, 1975 Man and Natural Resources Croom Helm; London.

Hill Stuart B and MacRae R.J. 1992. Organic farming in Canada. Agriculture, Ecosystem and Environment 39:71-84.

Hodges, R.D. 1983 Soil Degradation. A Global Problem. Soil Association Quarterly Review September, 1983, pp. 2-5.

Honik, S.B., L J. Sikora, S.B. Sterrrett, J.J. Murray, P.D. Millner, W.O. Burgae, D. Colacicco, J.F. Parr, R.L. Chaney, and G. B. Willson. 1984. Utilization of sew-age sludge compost as a soil conditioner and fertilizer for plant growth. Agriculture Information Bulletin No. 464. U.S. Department of Agriculture, U.S. GPO, Washington, DC. 32pp.

Howard A. 1940. Agricultural Testaments. The Oxford University Press. pp 233.

Howard, A. 1943. An agricultural testament. Oxford University press. New York.

Howard, A. 1947. The Soil and Health: a study of organic agriculture. Devin-Adair, New York. 253 pp.

Howard, A. and Howard, G.L.C. 1929. The application of science to crop production, and Experiment carried out at the Institute of Plant Industry, Indore. Oxford University Press, Bombay.

Howard, A. and Wad, Y.I. 1931. The waste products of agriculture their utilization as humus. Oxford University press, London.

Hynes, H. P. 1987. The recurring silent spring pergamon, New York.

I.I.R.D. 2001. Concepts, Principles and Basic Standards of Indian Organic Agriculture, Volume I, Institute for Integrated Rural Development. Nakshatrawadi, Aurangabad. p.p. 29.

IFOAM, 1996. Concepts, Principles and Basic standards of organic agriculture, Intern. Federation of organic agriculture Movements p.p. 13-14.

In: T. Edens, C. Fridgen, and S. Batttenfield (Editors), Sustainable Agriculture and Integrated Farming Systems. Michigan State Univ. Pr., East Lansing, MI. Pp. 84-95.

Ingle, H.V., Athwale. R.B., Ghawde, S.M. and Shivankar, S.K. 2001. Integrated nutrient management in acid lime. *South Indian Hort., 49:126-127.*

Ingle, H.V., Athwale. R.B., Ghawde, S.M. and Shivankar, S.K. 2001. Integrated nutrient management in acid lime. South Indian Hort., 49:126-127.

Jadhav, A.S., Verma, O.P.S., Shrinivas, S., Sheikh, A.A. and Harinarayana, G. 1992. Fert. News 37 (II) : 37-41.

Jeevan Rao, K. and Shantharam, M.V. 1995. Nutrient changes in agricultural soils due to application of garbage. Indian Journal of Environmental Health 37, 265-271.

Jenny Hand and Raychoudhary, S.P. 1960. Effect of climate and cultivation on nitrogen and organic matter reserves of Indian soils. Indian council of Agricultural Research, New Delhi Jones, M.J. 1973. J. Soil science, 24 :42.

Jha, Manoj, K. 2003. Organic farming. Way to food security. *Chronicle,* November, 2003, pp 33-34.

Jha, Manoj, K. 2003. Organic farming. Way to food security. Chronicle, November, 2003. pp 33-34.

Kachhave, K.G. 2002. Recycling of Agricultural and agro--industrial wastes and evaluation of their nutrient potential for sustainable agricultural production in soil research inventory of Marathwada Published by Dept. Agril. Chem. and Soil Science, MAU, Parbhani. pp 102-112.

Kalisko E. and L. Kalisko, Kalisko Archives, .1939.Agriculture of Tomorrow. London 1939.

Katyal, J.C. 2001. Fertilizer use situation in India. Journal of the Indian Society of Soil Science 49, 570-582.

Katyal, J.E. and B.D. Sharma. 1991. Geoderma, 49: 165. Lal, R. and B.T. Kang. 1982. Management of organic matter in Soils of the Tropics and Subtropics, Trans. 12th Intern. Congo Soil Sci. New Delhi. pp. 152-178.

Katyal, J.E., N.H. Rao and M.N. Reddy. 2000. Critical aspects of organic management in the tropics -Example - India J. Nutrient cycling Agro systems.

King, F.H. 1911. Farmers of forty centuries. Harcourt, Brace Publ. New York.

Koepf, H. 1989. The Biodynamic Farm. Anthroposophic Press, Hudson, NY. 245 pp.

Kramer, D. 1984. Problems facing Canadian farmers using organic methods. In: T. Schreker and R. Vleg (Editors). Pesticide Policy: the environmental imperative. Friends of the Earth, Ottawa, ON. Pp. 129-162.

Krauss, A. 2001. Balanced nutrient management in quality production. Proc. National Symposium on balanced nutrition of ground nut and other field crops grown in calcareous soils of India. Held at GAD, Junagadh, Sept. 19-20.2001, pp 171-183.

Krauss, A. 2001. Balanced nutrient management in quality production. Proc. National Symposium on balanced nutrition of groundnut and other field crops grown in calcareous soils of India. Held at GAU, Junagadh. Sept. 19-20.200 I, pp 171-183.

Kumaraswamy, K. 2000. Environment-friendly recycling of organic wastes. Kisan World (Nov.): 37-39

Kumarswamy, K. 2000. Organic farming - Relevance and prospects. Indian Soc. Soil Sci., Newsletter; 12:1.

Kurumthottical, Sam T. 1995. Assessment of phosphatic sources for possible heavy metal contamination and their bioavailability. Ph.D. Thesis, Post Graduate School, Indian Agricultural Research Institute, New Delhi. pp. 124 + x.

La Mondia, J.A., Elmer, W.H. Mervosh, T.L., Cowles, R.S. 2002. Integrated management of strawberry by rotation and intercropping. Crop Protection, 21 (9): 837-846.

Lady Eve Balfour .Guide to Organic Farming The Lifeblood of the Earth

Lampkin N. 1992. Rotation design for organic systems. Organic farming pp. 125-160 ipswick. U.K.

Lampkin, 1985. Biological farming systems in Europe.

Lin Bao and Tiwari, K.N. 2003. Personal Communication.

Lowdermilk, W.C. (1953). Conquest of the land through 7000 years. Agricultural Information Bulletin No. 99, U.S. Department of Agriculture.

MacRae, R.J., Hill, S.B., Henning, J. and Mehuys, G.R. 1989. Agricultural Science and Sustainable Agriculture: a Review of the Existing Scientific Barriers to Sustainable Food Production and Potential Solutions. Biol. Agric. Hortic. 6(3): 173-219.

MacRae, R.J. and Hill, S.B. 1990. Literature review: studies on the marketing of organic produce in North America and Europe. Report to the International Programs Branch, Agriculture Canada. Canadian Organic Growers, Ottawa.

MacRae, R.J., Henning, J. and Hill, S.B. 1990a. Strategies to overcome barriers to the developmenl uf sustainable agriculture in Canada: the role of agribusiness. Ecological Agriculture Projects Research Paper. Ste-Anne de Bellevue, QC.

MacRae, R.J., Hill, S.B., Henning, J. and Bentley, A.J. 1990a. Policies, programs .and regulations to support the transition to sustainable agriculture in Canada. American J. Alternative Agriculture 5(2):76-92.

MacRae, R.J., Hill, S.B., Mehuys, G.R. and Henning, J. 1990b. Farm-scale agronomic and economic transition to sustainable agriculture. Advances in Agronomy 43: 1.55-198.

Madden J. P. and Dobbs, T.L. 1990. The role of economics in achieving low input farming systems. In: C.A. Edwards, R. Lal, P. Madden, R.H. Miller and G. House (Editors), Sustainable Agriculture Systems Soil & Water Conservation Society, Ankeny, IA, pp. 459-477.

Malewar, G.U., Syed Ismail and Waikar, S.L. 2001. Indigenous nutrient management technology in selected agro-eco regions of Maharashtra. In: Indigenous Nutrient Management Practices: Wisdom Alive in India (Eds. Acharya C.L, Ghosh, P.K. and Subbarao, A.). Indian Institute of Soil Science, Bhopal, pp 55-70.

Malewar, G.U. 2004. Organic Manures: Concept and Future In a book "Ekatmic Khat Vyavasthapan" p.p.45-54.

Malewar, G.U. 2004. Soil organic matter: A fundamental soil quality indicator in sustainable Farming, State level Sem. Rahuri chapter Indian Soc. Soil Sci. MPKV, Rahuri, Theme paper.

Malewar, G.U., A.R. Hasnabade and Syed Ismail. 1999. J. Maharashtra Agric. Univ. 24(2): 121-124.

Malewar, G.U., P.P. Ramaswami and L. Sushila Oevi. 1998. Bull. Indian Soc. Soil Sci. 19: 48-57.

Malewar, G.V., Syed Ismail and Rudraksha, G.B. 1998. Integrated nitrogen management in chilli. In: Integrated plant nutrient supply system for sustainable produl:tivity (Eds. Acharya C.L. et al.,), Indian Institute of Soil Sl:ienl:e, Bhopal, pp 156-163.

Manna, M.E. 2002. Long term effects or fertilizers and manures on soil organic Pools and Sequestration under different cropping systems in sustainable agriculture winter school, INM for sustainable agriculture, Dept. of ACSS, or. POKY, Akola. p.p. 275-279.

Mathur, B.S., Sarkar, A.K., Singh, K.P. and Lal, S. 1989. Soil Science and Agricultural Chemistry, BAU. Re-search Bulletin No.2: Birsa Agricultural University, Ranchi.

Menon, T.G.K. and V.B. Karamarkar. 1994a. Biodynamic agriculture. In: Organic Farming (Y.N. Shroff et al., Eds.) Jawaharlal Nehru Agricultural University, Jabalpur. pp. I 16-I 18.

Menon, T.G.K. and V.B. Karamarkar. 1994b. The planting calendar and its e Land-water Interface. (K. Steele Ed.) CRC Press Lewis Publishers, Boca Raton Fl. pp. 57-68.

Mikkelsen, R.L. and Gilliam, J.N. 1995. Animal waste management and edge field losses. In: Animal Waste and the Land-water Interface. (K. Steele Ed.) CRC Press Lewis Publishers, boca Raton Fl. pp. 57-68.

Mishra, B., A. Sharma and A.K. Sarkar. 1992. Efficiency of organic manures for Groundnut-wheat sequence in Acid Alfisol of Ranchi. Proc. Natn. Sem. Organic Farming, MPKV, Pune, P.P. 1-3.

Mishra, M.M., Khurana, A.L., Dahia, S. S. and Kapoor, K.K. 1984. Phospho-compost. Trop. Agric. 61: 174.

Mishra, R.Y. and Hesse, P.R. 1982 Comparative analysis of organic manures project document No. 24 FAO/ UNDP Rome.

Mollison, W. 1988. Permaculture : A Designer's Manual. Tagari Books, Tyalgum, Tasmania, Australia. pp. 576.

Motsara, M.R. 1993. National project on Biofertilizer -status position VIIIth plan proposal. Nat. Conf. Biofertilizers and organic farming. Ministry of Agriculture, Govt. of India and Oept. of Agriculture, Govt. Tamilnadu, Madras. P.P. 14-20.

Mutsuaki, T. 1977. Studies on utilization of animal wastes in agriculture. Bull. Agric. Res. Insti. Kangana Prof. Japan. pp. 118.

Narvwal, S.S. and Iswarsingh. 1995. Synergism in crop-ping system. Organic Agriculture, edited by P.K. Thompson 173-192.

Nawale, A.R. 1998. Integrated nutrient management for sapota *(Achras zapata* L.) Ph.D. Thesis, MAU, Parbhani.

Nirmala, L. 2003. Best quality produce through organic farming. Kisan World, 30(12):30-31.

Northbourne, W.E. 1940. Look to the land Dent, London.

Palaniappan, S.P. and Annadurai, K. 1999. Organic farming: Theory and Practice Pub, Scientific publishers (India). Jodhpur.

Palaniappan, S.P. and Annadurai K (1999) Organic Farm-ing - Theory and Practice. Scientific Publishers (India) Jodhpur.

Paroda R.S. 1997 Key note address in FAO-IFFCO Inter-national seminar on lPN'S for sustainable devel-opment Nov. 25-27 Vidgyan Bhavan New Delhi.

Parr, J.F., Papendick, R.I. and Colacicco, D. 1986. Recycling of organic wastes for a sustainable agriculture. Bio. Ag. Hort. 3: 1 15-130.

Pathak, R.K. and Ram K.A. 2003. Biodynamic Agriculture Bulletin No. 14, Central Institute for subtropical Horticulture, Lucknow. pp 42.

Patil, A.J. and S.O. Kulkarni. 1998. Effect of organic recycling of Subabul on yield, nutrient uptake and moisture utilization by Rabi sorghum. Abst. P.P. 30-3 I.

Patil, B.H., P.H. Rasal and P.L. Patil. 1992. A study on combined effect of biofertilizers on nodulation and Biomass production of cowpea. Proc. Nat. Sem. organic fanning. Mahatma Phule Krishi Vidyapeeth, College of Agriculture, Pune P.P. 99-100.

Patriquin, D.G., Hill, N.M., Baines, D., Bishop, M., and Allen, G. 1986. Observations on a mixed farm during the transition to biological husbandry. Biol. Agric. Hort. 4:69-154.

Pieters, A.J. 1927. Green manuring. Wiley, New York.

Pugh, T. (Editor). 1987. Fighting the Farm Crisis. Fifth House, Saskatoon, SK. 129 pp.

Purakaystha, T.J. and Chhonkar, P.K. 2001. Influence of vesicular arbuscular mycorrhizal fungi (*Glomus etllllicatum* L.) on mobilization of zinc in wetland rice (*Oryza sativa* L.) Biology and Fertilit)' of Soils 33, 323-327.

Rajendra Prasad 2000. Nutrient management strategies for the next decad3-18. race element composi-tion of fertilizers and soil amendments. Journal of Environmental Quality 26, 551-557.

Rajendra Prasad and Power, J.F. 991 Crop residue management. Adv Soil Sci. is 205-249.

Rajendran, T.P., Venugopalan, M.Y. and Tarhalkar. P.P., 2000. Organic Cotton Farming in India. CICR Technical Bulletin No.1. 39 p + X. Central Institute for Cotton Research, Nagpur

Ramaswamy, P.P. 1999. Recycling of Agricultural and agro-industrial wastes for sustainable agricultural production. J. Indian Soc. Soil Sci. 47 (4): 661-665.

Raut, R.S., G.U. Malewar and A.B. More. 1995. In: A book "Inoculation in production of oil seeds". AICRP on Biological N fixation. oept. of Agril. Chem. Soil Sci., Marathwada Agril. University, Parbhani.

Reddy. G.R., G.U. Malewar, K.L. Sharma, N. Sudhakar and B.G. Karle. 2001. Effect of organic recycling on soil improvement and rainfed crop production in Vertisol Indian J. Dryland Agric. Res. & Dev. 16(2): 91-96

Report of Technical Team on organic Farming 1994. Ministry of Agriculture Department of Agriculture and co-operation of Agriculture and Co operation, Govt. of India.

Robinson, P. 1985 Effects of a transition to ecological-, organic agriculture on livestock production in Manitoba. M.Sc. Thesis. University of Manitoba, Winnipeg, MN. pp. 101.

Robinson, P. 1986. Searching for alternative solutions: sustainable agriculture. Policy Branch Working Paper, Agriculture Canada, Ottawa, ON. pp. 18.

Rodale, J.R. 1948. The healthy Humus, Rodale Press, Emmaus, Pa.

Rudolf Hauschka. 1987. The Nature of Substances, Published by Vincent Stuart Ltd. London

Rudolf Steiner .1974.Agriculture .Published by R. Steiner Press London 1974.

Rynk, R. 1992. On farm composting hand book. NRAES-54, Ithaca, Natural Resource, Agri and Engg. Ser. Co-op Extn. 1-186.

Sale, K. 1985. Dwellers in the Land: the bioregional vision. Sierra Club Books, San Francisco. pp. 217.

Sankaran, A. 1996. Soil fertility management for reconcilling sustainability with productivity. J. Indian Soc. Soil Sci., 44 (4):593-600.

Sankaran, S. 1993. Potential of Biofertilizers in Indian Agriculture, Nat. Conf, Biofertilizers and organic farming. Ministry of Agriculture, Dept. of Agriculture and cooperation. Govt. of India and Dept. of Agriculture. Govt. of Tamilnadu. Madras. P.P 5-9.

Sansavini, S. and J. Wollesen, 1992. The organic Farming movement in Europe. Proceedings of the workshop on History of the organic movement held at the 88 the ASHS Annual meeting the Pennsylvania State University, University Park U.S.A. on 24-July-1991.

Schuphan Werner, 1974. Nutritional value of crops as in-fluenced by organic and inorganic fertilizer treat-ments: Results of 12 years experiments with veg-etables (1960-1972). Qual. Plan Foods Hum. Nutr. 23, 333-358.

Schuphan Werner, 1974. Nutritional value of crops as influenced by organic and inorganic fertilizer treatments: Results of 12 years experiments with vegetables (1960-1972). Qual. Plan Fds. Hum. Nutr. 23, 333-358.

Science Council of Canada, 1986. A growing concern: soil degradation in Canada. pp. 24.

Sekhon, G.S. 1997. Nutrient needs of irrigated food grain crops in India and related issues. In: Plant nutrient needs, supply, efficiency and policy, issues 2000-2025 (J.S. Kanwar and J.C. Katyal, Eds.) National Academy of Agricultural Sciences, New Delhi. pp 78-90.

Senate of Canada, 1984. Soil at Risk. Supply and Service Canada, Ottawa, ON. pp. 129.

Sharma, A.K. 2001. Quality of organic product. In: A Handbook of Organic Farming. Agrobios (India), Jodhpur, pp 442-443.

Sharma, N.K. 1995. Comparative study on the effect of biofertilizers, compost and chemical fertilizers on the growth, yield and quality of okra. M.Sc. (Agri.) Thesis, JNKVV, Jabalpur.

Shea, K.P. 1973. A celebration of silent spring Environment 15:4-5.

Shinde, P.H. and D.B. Gawade. 1992. Effect of application of Farm Yard Manure on the availability of NPK & B in soils. Proc. Natn. Sem. Organic Fanning. Mahatma Phule Krishi Vidyapeeth, Pune. P.P. 10-11.

Shinde, V.S., F.R. Khan and C.D. Mayee. 1996. Ann. Rept. Cropping systems and verification function, NARP, Aurangabad,

Shreshtha, Y.H., Ishii, T., Matsumoto, I. and Kadoya, K. 1996. Effect of VAM fungi on Satsuma mandarin tree growth, water stress tolerance, fruit development and quality. J. Japanese Soc. Hart. Sci., 64(6):801-807.

Singh, Bijay and Singh, Yadvender. 1997. Green manuring and biological N fixation: North Indian perspective. Plant Nutrient Needs, Suply, Efficiency and Policy Issues: 2000-2025. (J.S. Kanwar and Katyal J.E., (Eds.) National Academy of Agricultural Sciences, New Delhi pp. 29-44.

Singh, T. 1993. Prospects and Potential of Biofertilizers a mission mode approach. Nal. Conf. Biofertilizers and organic farming. Ministry of Agriculture, Govt. Of India and Dept. Of Agri. Tamilnadu, Madras. PP. 10-13.

Stchouwer, R. 1999. Land application of sewage sludge in Pennsylvania: What ails sewage sludge and what can be done with it? http://www.agronomypsu.edu/extension/facts whatis-pdf.

Strauss, M. 1990. Green shoppers are a force to reckon with. Globe and Mail. 12 April.

Swaminathan, M.S. 2003. Enhancing our Agricultural Competitiveness. 6th J.R.D. Tata Memorial Lecture 26'h August, 2003, ASSOCHAM, New Delhi.

Swarup, D.O., Damodar Reddy and R.N. Prasad. 2001. Long term soil fertility management through IPNS. Indian Institute of Soil Science Bhopal, Indian.

Tachibana, S. and Yahata, S. 1998. Effects of organic matter and nitrogen fertilizer application for a high density planting of Satsuma mandarin. J. Japanese Soc. Hart. Sci., 65(3):471-477.

Taillefer. D. 1989. Organic food supply and demand in the National Capital Region. Friends of the Earth, Ottawa.

Tandon, H.L.S. (Ed.). 1994. Fertilisers, organic manures, recyclable wastes and biofertilisers - components of integrated plant nutrientsFertiliser Development and

Tandon, H.L.S. 1997. In Plant Nutrient Needs Efficiency and Policy Issue-2000-2025. National Academy of Agricultural Sciences, New Delhi. P.P. 15-28,

Thampan P.K.I 995.0rganic Agriculture. 1995. Peekay Tree Crops Development

Theodora Stew .1976.Sensitive Chaos, Published by Stockmen Books N.Y. 1976.

Theriault, J. 1988. La Situation et de Developpement de)' Agriculture Ecologique au Quebec. Ministere de l' Agriculture, des Pecheries et de l' Alimentation du Quebec. 111 pp.

United State Department of Agriculture. 1987. Improving Soil with Organic Wastes. Report to the Congress in response to Section 1461 of the Food and Agricul-ture Act of 1977 (PL 95-113). U.S. GPO, Washing-ton, DC. 157 pp.

Vachon, D. 1988. Presentation a la Journee production vegetale et sols. Compte Rendu de la Semaine de l' agriculture, de l' alimentation et de la consommation. Universite Laval, Quebec. 17-21 mars.

Vioayakumar, K.R., Mammean, G., Pillai, G.G. and Vamadevan, Y.K. 1986. Alley cropping of leucaena in coconut gardens in Western Ghats of India.

Virmani, S.M., K.L. Sahrawat and J.R. Burford. 1982. In Vertisol and rice soils of the tropics.Trans.12Th Inter. Congress. Soil Sci., New Delhi, PP. 80-93

Wagstaff, H. 1987. Husbandry, methods and farm systems in industrialized countries which use lower levels of external inputs: a review. Agriculture. Ecosystems and Environment. 19: 1-27.

Walker, J.L. and Walker, L.J. S. 1988. Self-reported stress symptoms in farmers. J. Clinical Psychology 44(1):10-16.

Walters, C. and Fenzau, C.J. 1979. An Acres USA Primer. Acres USA, Raytown, MO. pp. 464.

Wigle, D.T., Semencier, R.M., Wilkins, K., Riedel, D., Ritter, L., Morrison, H.I., and Mao, Y. 1990. Lymphoma mortality and agricultural practices in Saskatchewan. J. National Cancer Institute 82:575-582.

Woese, K.D., Lange, C.B. and Bogel, K.W. 1997. A comparison of organically and conventionally grown foods - Results of a review of relevant literature. Journal of science, Food and Agriculture 74: 281-293.

Yield of dry matter and organic nitrogen. Leucaena Research Reports, 7:72-74.

Zende, G.K. 1998. Factors influencing organic agriculture. INORA, News Letter. 2(2): 1-2.

Zhao, S.P., Wang, M., Zhang, L., Zhou, J., Lei, P. F. 2002. An ecological approach to establishment of citrus orchards with reference to pollution free horticultural technique for growing citrus tree. A case study in Dongjang Reservoir area, Acta Agriculture Universitatis- Jiangxiensis, 24 (5): 661-666.

Ziauddip, S. 2000. Integrated nutrient management in banana cv. Ardhapuri (Musa AAA group Cavendish Subgroup). Ph.D. Thesis, MAU, Parbhani.